# AGENT
# SMARTS

Website
IDX Integration
Social Media
SEO
Content Strategy

Leveraging WordPress to create an online presence
while engaging your community and growing your **REAL ESTATE** business

## Marna Friedman

Agent Smarts: Real Estate Websites Made With WordPress
Published by MWF Publishing, Dallas, GA

**Other books by Marna Friedman**
EVENTually Perfect
The Small Business That Could™
The Small Business That Could™ For Women
The Social Launch Toolkit
Real Smarts: Leveraging smartphones and new technologies in your Real Estate Business

ISBN: 978-0-9840169-2-1
Library of Congress Control Number: 2013913972

Manufactured in the United States of America

# Table of Contents

# Symbol Key:

There are several items that I wanted to call out in this book. They are each broken out into specific types of categories:

Notes on technology

Definitions

Real Estate Detail

# Foreword

I find it interesting that so many people think that because Word-Press is an open source platform available for free, it must be easy. And in so many ways it is. But designing a professional website that represents you and includes all of the functionality you may need and/or want can be an overwhelming and frustrating task.

As a real estate professional, you probably have very strong feelings about FSBOs and why a homeowner would probably have better results using a professional. Yet many real estate professionals don't feel that way about their websites. So let me point out some things:

- There are already several how-to books available on using WordPress. Unlike those, this book focuses on what's possible with WordPress in building a **real estate website**. It provides tips and tricks that will help you see what's possible.

- Most professional website developers (*myself included*) have received training in web design, graphic design and programming, and continue to learn in order to provide the best service to our clients.

- While WordPress itself is free, as are some themes and several plugins, much like a FSBO, a website can usually look better and offer a better user experience when developed by a professional.

- Even though you may end up hiring a professional to build your website, this book seeks to provide you with the knowledge of what's important for a real estate professional and what's possible with WordPress.

While I was a website developer before I became a real estate agent, I had to go to school to learn about real estate. A book is great to learn what's possible, but nothing compares to actually doing something. So step lightly, learn what's possible, and then decide whether you can actually do it yourself, like a FSBO, or if you are looking to achieve the most value by hiring a professional.

*Marna Friedman*

http://mwfmedia.com
marna@mwfmedia.com

# Why WordPress

Learn how to create a WordPress website
    📖 WordPress as a CMS
    📖 Advantages of WordPress

## WordPress as a CMS

While there are many options available for your website platform, a content management system (CMS) allows you to easily update your content. WordPress also enables you to quickly launch a website for your real estate business and plugins help you expand the functionality of your website.

NOTE: *unlocking a technology tidbit for you*

*A CMS like WordPress enables you to own and control your website, content and links.*

NOTES

_____

_____

_____

_____

_____

Most real estate professionals have a website, and many of them are paying dearly for them.

It's no secret that the real estate market has moved online. Last year, over 90% of home buyers started their search for a new home on the internet, while over 70% used the same medium to find an agent. With these kind of statistics, it's vital to have an effective, professional real estate website to compete.

Most major real estate firms offer their agents a website. This can be a quick and easy way for a real estate agent to establish themselves online. But since agents are independent contractors who may not stay with the same broker, they need to understand that they can't take that website with them when they transfer to another brokerage.

Another issue for real estate agents is the learning curve. Most "packaged" real estate websites, whether from a broker or purchased through another company typically promise training and support, which to an agent struggling to find extra time, can be quite appealing. But this too, comes with issues. The agent needs to learn the platform and when they can't find time to accomplish this, the site usually just becomes a listing portal, not unlike many others.

Ask your typical real estate agent about WordPress, and most won't know what you're talking about. If they have heard of WordPress, chances are they think it's simply a blogging platform, maybe something to use in conjunction with their main site, if at all. However, WordPress offers affordability, functionality and power to provide a comprehensive, all-in-one real estate solution for even the most demanding of agents.

NOTES

_____

_____

_____

_____

_____

It's important to understand how a content management system can help with your real estate business strategy before, during and after your sales and/or listings. And using WordPress as your CMS offers tremendous possibilities. According to Forbes magazine, in 2012 WordPress powered one of every 6 websites on the Internet, nearly 60 million in all, with 100,000 more popping up each day. And today, according to w3techs, WordPress is used by 59.1% of all the websites whose content management system we know. This is 20.4% of all websites.

# Advantages of WordPress
There are several reasons why WordPress continues to be the CMS of choice, but the major reasons are:

### SEO
Everyone wants their website to be found by search engines, and they frequently struggle with search engine optimization, sometimes spending a significant amount of money on it. But WordPress offers clean code that is properly formatted and organized in a simple way, as well as offering SEO friendly URLs that enable every page to have a distinctive title and meta description.

### Easy Set-up and Launch
WordPress was developed to be easy, and once installed can be used immediately. For those that are hesitant, they can practice with a WordPress.com website and learn the basics before launching a self-hosted site. And then if they want something more unique, it is quite easy to find a developer to help with that. A system as popular as WordPress offers access to help,

NOTES

_____

_____

_____

_____

_____

whether it be local groups, hiring a developer, or seeking more formal education.

### Themes

Another plus related to the widespread affinity for WordPress is that custom, unique templates are easy to find and sometimes free. There are also an endless number of independent designers out there who create SEO friendly templates.

### Options to Optimize

Adding plugins to your site is helpful if you're looking to increase its functionality. Before making any changes, ensure that they're SEO friendly and will help instead of hurt when it comes to local search.

# In the Know

Many people will tell you how easy WordPress is, and compared to many other web platforms, this is true. But setting up a website may not be your thing. Be mindful of what you do best and remember why people choose you as their agent. It's OK to be good at some things and not others. Knowing your strengths is great. Knowing when to seek help is even better. Explore  what's possible and then make an educated decision.

NOTES

_____

_____

_____

_____

_____

# Website Basics

**Getting Started**
- 📖 Naming your website
- 📖 Hosting your website

# Getting started

Before you can install WordPress, you will need to have a domain name and hosting service. If you already have a registered domain and a hosting service, you should find out from them about adding WordPress to your current site, and read through this chapter before installing. The questions you should ask them include:

- Does your host offer WordPress?
- Can you *add* WordPress to your current site?
- Can you redirect the WordPress URL to your current website?
- Can you just add a page to your current site for your WordPress blog?

We'll answer these questions and more when we go over hosting later in this chapter, but let's start with your name.

NOTES

_____

_____

_____

_____

_____

_____

# Naming Your Website

You have several choices when it comes to the name of your website (domain):

- It *can* be your name or the name of your business
- It *should* be a name that depicts your real estate territory or niche
- It *should* be short and easy to remember
- It *should* be easy to spell - it will help people find it.
- It *should* be a .com - if someone else already has the .com, and you go with a .net (or something else), guess who will get found first? Most of us just "default" to .com, so keep this in mind when naming your website - especially before you start printing business cards and other marketing materials.

**Real Estate Domain Names:**

*Most real estate agents use their name for their website, including myself, but I have learned that may not be the best idea.*

The more specific your website name, the more people will be able to clearly understand what you offer. So if you are selling real estate in Palm Beach, palmbeachcondos.com might be good, but downtownpalmbeachcondos.com is better. It clearly states that you specialize in downtown Palm Beach condos. You can provide landing pages for each condominium building with information about it and listings. And if you want some

NOTES

_____

_____

_____

_____

_____

SEO help, you can also purchase downtown-palm-beach-condos.com and point it to your downtownpalmbeachcondos.com. This can also help you *own* that brand, since you will own all of the ways the domain name might be used.

NOTE: *unlocking a technology tidbit for you*

### Trademark

*If you create a unique name, or find the "perfect" name, check to make sure it is not trademarked. Before you spend money on logo design, marketing and more, check to see if your name is trademarked. And if you think it is a truly great name, you might want to consider trademarking it yourself.*

So, if you have already registered your domain, you can skip this step. If not, here are some articles that can help you understand the importance of a domain name:

- ☐ **http://www.sitepoint.com/tips-c**
- ☐ **http://moz.com/blog/how-to-choose-the-right-domain-name**
- ☐ **http://www.huffingtonpost.com/tom-lowery/5-tips-for-choosing-a-the_b_3859497.html**

For help in identifying a domain name:

- ☐ **Nameboy**
  **http://www.nameboy.com/**
  Search for a domain name by inputting keywords or terms that you would like in your domain name.

NOTES

_____

_____

_____

_____

_____

_____

NameBoy will generate domain names based on the keywords you enter.

☐ **Dot-o-mator**
**http://www.dotomator.com/**
Search terms (or groups) of terms you would like to use in your domain, and this application searches the availability for you.

You can also search your area and identify keywords that are frequently used. These might help you with naming your website. But don't try to be so descriptive that you end up with a very long name or too many names and no real identity. The goal is to find a middle ground that helps identify your territory, while helping brand your business.

# Owning Your Name

Once you decide on your domain name, you need to purchase it. Some host companies will offer you a free domain when you purchase your hosting from them. Don't do it! A domain name costs under $20/year. If you ever decide to switch your host provider, you may lose that domain.

Real Estate Detail

**Ownership**
*As a real estate professional, you understand the importance of ownership, do not "rent" your website or use a "free" address.*

NOTES

_____

_____

_____

_____

_____

So where can you purchase a domain? You can purchase a domain through a registered domain provider which includes some host providers or you can purchase separately and forward to your host. Some things you want to make sure of:

- Ensure you have a right to the name. Check for trademark and also check for cybersquatting.

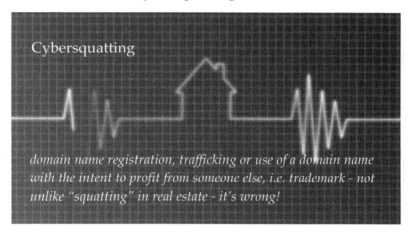

Cybersquatting

*domain name registration, trafficking or use of a domain name with the intent to profit from someone else, i.e. trademark - not unlike "squatting" in real estate - it's wrong!*

- If you are buying a domain name from auction, consult with an attorney to ensure that you have a legal right to the name.

- If a name you want is unavailable, create a "watch" so that if it becomes available, you can purchase it at a later date.

- Check to see if the name you want has been used before and appears in search results. Could this be detrimental to your business?

NOTES

_____

_____

_____

_____

_____

_____

- Check to see what security features your domain provider offers including:
  - Transfer Lock
  - Name Safe
  - Forwarding
  - Email
  - DNS Management

# Hosting your website

If you purchase your domain elsewhere, you need to make sure that your domain "points" to your hosting service. You also need to make sure that your hosting provider offers WordPress hosting. There are hosting companies that offer WordPress packages as well as companies that only offer WordPress hosting.

Some web developers may offer to host your site and manage it for you. The caveat here is to ensure that you own your domain name so that if you ever decide to end that relationship, you can do so without losing your domain name and search traffic you have established. You should also review the contract to understand what your rights are when you end the relationship.

Another factor is to understand what type of control panel your host provides. Many web developers charge extra for using host providers that offer proprietary hosting. This is because there is a learning curve in setting up the website and many times the

NOTES

_____

_____

_____

_____

_____

particular host provider does not offer the support needed.

**cPanel**
*Unix based web hosting control panel that is somewhat of an "industry standard" offered by several host providers*

Some hosting tips:

- Check with your web developer to find out what host providers they prefer to work with. They should offer you a variety for you to select from, or they may just recommend that you select one that uses the Unix control panel - but it's always a good idea to confirm with them before you purchase hosting service.

- Since you may want to have "landing page" sites for marketing efforts, make sure your hosting plan allows for unlimited domains and subdomains. You should not need to purchase additional hosting plans for these, but this may mean you need to purchase a more robust plan on some hosts.

- Make sure that your host provider offers the most recent version of PHP. While WordPress will run on PHP 5.2.4 or greater, several frameworks (themes) require 5.4 or greater, and the most recent stable version is 5.5.6.

Notes

- Search social media and the internet for feedback on the host provider. If there are negative comments, research them before you make a decision.

- Make sure that the hosting company offers the minimum requirements set by WordPress:
  - PHP version 5.2.4 or greater
  - MySQL version 5.0 or greater
  - mod_rewrite Apache module
  - *suPHP is an added benefit for security purposes*

Here are some host providers we recommend:
- BLUEHOST
  http://www.bluehost.com/

- HOSTGATOR
  http://www.hostgator.com/

- HOSTMONSTER
  http://www.hostmonster.com/

- HUB
  http://www.webhostinghub.com/

- JUSTHOST
  http://www.justhost.com/

- SITE 5
  http://www.site5.com/

NOTES

_____

_____

_____

_____

_____

_____

# Managed WordPress host providers

As the title implies, managed WordPress hosting is specifically for WordPress. These hosts usually provide update support and more. However, some of these providers also limit the plugins you can use on your website. So research what you can and can't do before making a selection.

- ☐ **KahunaHost**
  **http://kahunahost.com/**
  Cloud hosting and Varnish Caching.

- ☐ **Lightningbase**
  **http://lightningbase.com/**
  Basic management, with per-site backups, caching/CDN setup and more.

- ☐ **Pagely**
  **https://pagely.com/**
  Pagely handles upgrades, performance, and security for you.

- ☐ **Synthesis**
  **http://websynthesis.com/**
  Hosting includes speed and security, as well as content marketing and SEO tools

- ☐ **WP Engine**
  **http://wpengine.com/**
  WPEngine emphasizes page load speeds and security.

NOTES

_____

_____

_____

_____

_____

_____

# Shared Hosting vs. Managed Hosting

There are pros and cons to either type of hosting, but here are some things to keep in mind:

| SHARED HOSTING PROS | SHARED HOSTING CONS |
| --- | --- |
| Less expensive | Usually older versions of PHP |
| Easier to use | May not know WordPress |

| MANAGED HOSTING PROS | MANAGED HOSTING CONS |
| --- | --- |
| Dedicated WordPress | Expensive |
| Performance | May charge per site |
| Security | Learning curve |
| Support - WP specific | |

# Domain Name Redirection

Once you have purchased your domain and hosting, you may need to do one more thing before you can install WordPress on your domain, and that is to redirect your domain name to your host provider. Depending on where you purchased your domain and host, they should provide you with directions or assist you in completing this.

You can redirect a domain name several ways, and here are two:
* Setting nameservers
    Your host company should provide you with two
    nameservers which are usually in numerical order and you
    will need to use these to redirect your domain to the host

NOTES

_____

_____

_____

_____

_____

- A name
  You might want to use this if you are using email from the company that you purchased your domain from, but want to host your site elsewhere. You will need to update your DNS Manager to the IP address provided by the host.

The information you will need from your host company:
- Host provider
- IP address
- Nameservers

You should make sure that you have the following information for future use and/or your web developer:
- Host name
- FTP access
- cPanel access
- Account pin if needed for support

NOTE: *unlocking a technology tidbit for you*

**Information:**
*Never email password or access information. Share it via a secure system only.*

Once you have your domain name set up on your host, you can install WordPress.

NOTES

_____

_____

_____

_____

_____

_____

## NOTES

_____

_____

_____

_____

_____

_____

# Getting Started

## Getting Started with WordPress
- 📖 WordPress.com vs. WordPress.org
- 📖 Installing WordPress
- 📖 Understanding the Dashboard

# WordPress.com vs. WordPress.org

WordPress offers two different platforms and only one is for self-hosted sites. WordPress.com (not including the VIP program offered by WordPress) offers free hosting for your website as well as several free themes and plugins. However, they are limited to what they offer, and not necessarily what you want or need. Here are some basic comparisons:

|  | WordPress.com | WordPress.org |
|---|---|---|
| Hosting | Free | You need to select provider |
| Themes | Free – limited | Free and Premium |
| Plugins | Only those built-in | Free and Premium |
| Design Modifications | None | Unlimited |

There are other issues, but I felt that these were the most important differences and strongest reasons to use WordPress.org.

NOTES

_____

_____

_____

_____

_____

_____

# Installing WordPress

There are several ways to install WordPress. You can determine which way is best for you. Once WordPress is installed, you can begin modifying the "look" of your site, expanding the functionality of your site with plugins, or adding content.

Before you begin the installation process, you need to make sure you have the following:

1. Access to your web server
   While you can use your own server including software such as Apache and operating system software such as Linux or Unix, this is typically done through a hosting service that already provides these items. While there are several hosting providers, WordPress recommends four companies (http://wordpress.org/hosting/).

   Usually the easiest way to access your web server is by typing your domain/cpanel or domain:2082 (domain:2083 if you are using an SSL certificate on your site).

2. A text editor
   Some of the files in your installation may need to be modified, and to do this you will need a text editor. You will also need a text editor if you want to modify your theme after it is installed.

NOTES

_____

_____

_____

_____

_____

_____

3. **An FTP Client**
   FTP (File Transfer Protocol) can be used to download or upload files to your site, especially since it is usually much faster than through your browser. The following FTP clients can help you with this.
   - ☐ **http://filezilla-project.org/**
   - ☐ **http://cyberduck.ch/ (Mac)**

Now that you have all of the tools you need, you can install WordPress. There are several different methods of installing WordPress and you can review each one before making a decision on which one you will use. Some of these methods include:

1. Famous 5-minute install
   Directions for this process can be found at http://codex.wordpress.org/Installing_WordPress#Detailed_Instructions

2. cPanel
   This is done through the hosting provider cPanel. Directions for this process can be found at http://codex.wordpress.org/Installing_WordPress#Using_cPanel

I could go into detail on installing WordPress, but the links above take you directly to the detailed instructions. I recommend either of these installations vs. the quick install and Fantastico installations for several reasons, but mostly for security.

NOTES

_____

_____

_____

_____

_____

_____

**Security Tip:**
*Do not use ADMIN for Admin access data. Select a name you will remember, but that is unique. The Admin nickname can be the same name or different if you choose.*

WordPress will be installed and you can begin working with your website.

# Activating WordPress

Once you have completed the installation process, you will need to activate your WordPress site.

Type your domainname.com/wp-admin in your browser and the following screen will appear:
* Type in your Site Title
* Type in your Username (*use something unique*)
* Type in your password - twice
* Type in your email

**Security Tip:**
*Use a secure and unique password. A password program can help you create one. See list on page 22.*

NOTES

_____

_____

_____

_____

_____

_____

## Welcome

Welcome to the famous five minute WordPress installation process! You may want to browse the ReadMe documentation at your leisure. Otherwise, just fill in the information below and you'll be on your way to using the most extendable and powerful personal publishing platform in the world.

## Information needed

Please provide the following information. Don't worry, you can always change these settings later.

| | |
|---|---|
| Site Title | [ ] |
| Username | |
| | Usernames can have only alphanumeric characters, spaces, underscores, hyphens, periods and the @ symbol. |
| Password, twice | |
| A password will be | |
| automatically generated for | |
| you if you leave this blank | |
| | Strength indicator |
| | Hint: The password should be at least seven characters long. To make it stronger, use upper and lower case letters, numbers and symbols like ! " ? $ % ^ & ). |
| Your E-mail | |
| | Double-check your email address before continuing. |
| Privacy | ☑ Allow search engines to index this site. |

[ Install WordPress ]

## NOTES

_____

_____

_____

_____

_____

_____

_____

The last option is to select whether to have your site visible. But this is an option, and may or may not be searched, regardless of whether you choose not to.

Click Install WordPress.

If you need help selecting a strong password, you can use a site like:

- **Strong Password Generator
  http://strongpasswordgenerator.com/**
  This website generates new passwords and you can choose the number of characters in your password, in addition to specifics including letters, numbers, capitals, symbols, etc.

- **LastPass
  https://lastpass.com**
  LastPass is a secure password keeper that you can access from your browser, tablet and smartphone. You can save your login information to LastPass and not have to remember any passwords again ... except for your LastPass master password, which you should never make available to anyone.

NOTES

_____

_____

_____

_____

_____

_____

# Understanding the Dashboard

Now that you have installed Word-Press, you should start exploring the dashboard and learning what's possible, and to set up your site. And while it seems strange, you are going to start at the bottom of the dashboard with the Settings.

Click on the Settings section and start with General Settings.

### GENERAL

The General Settings control the basic functions of your site including title, date and time. While you can change these at any time, you should set them when you first start using your website.

Some of the features you can set here include the time zone you are located in, date format that appears on your site, the email you will use for website notifications and much more. All of these features can be modified at a later date as well, but it is wise to make sure that you set these properly

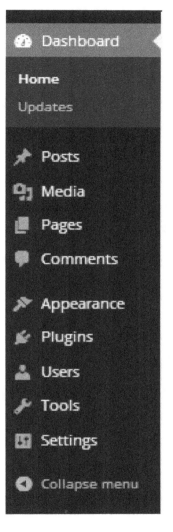

NOTES

_____

_____

_____

_____

_____

_____

right from the start.

NOTE: The screen captures in this book are from the most recent version of WordPress 3.8, and if you have not updated your site, the screens and options may be slightly
different.

# General

The first section of your Settings is where you can set the date
and time of your website. This will effect your website posts,
etc.

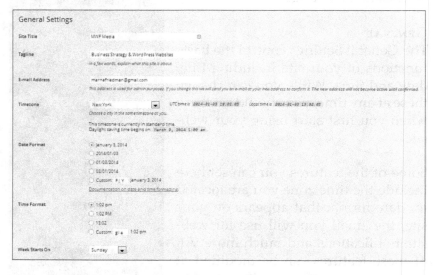

Note that the default day of the week is Monday. I change this
to Sunday, but you can leave it as is.

NOTES

# Writing

Formatting     The first option turns emoticons into symbols when checked. The second option will fix invalid XHTML code when checked

Default Post Category  The default category for blog posts if you don't assign a category

Default Link     Same as posts, but for your Links

Press This     A "bookmarklet" that can be placed in your browser bookmark bar so you can "grab" bits of content from the web. This feature can help with content creation

# Reading

These settings control how your site appears to your readers.

Front Page Displays:

Front Page    If you want your home page to look like a "website" and not a blog, you can create a landing page to appear on your home page and designate another page to show your blog posts

Blog Pages Show  You can determine how many blog posts you want on your blog page. The remaining posts will be visible through your archives.

NOTES
_____
_____
_____
_____
_____
_____

## Reading Settings

| Front page displays | ○ Your latest posts |
| --- | --- |
| | ⊙ A static page (select below) |

| | Front page: | Home ▼ |
| --- | --- | --- |
| | Posts page: | Blog ▼ |

Blog pages show at most    10 ⬍ posts

Syndication feeds show the    10 ⬍ items
most recent

For each article in a feed, show    ○ Full text
⊙ Summary

Search Engine Visibility    ☐ Discourage search engines from indexing this site
*It is up to search engines to honor this request.*

[ Save Changes ]

| Syndication Feeds | Similar to Blog Pages, you can determine how many posts a reader can see in your RSS Feed |
| --- | --- |
| Feed Articles | This determines what is displayed in a feed. |
| Search Engine | Select if you want to discourage site engines from indexing your site (while setting it up) |

## NOTES

_____

_____

_____

_____

_____

_____

# Discussion

This is the page that enables you to determine how you want your readers to interact with you when they comment on your site. This is an important part of setting up your website to prevent spamming. One of these items are particularly important:

## Other comment settings

Statement 2 - Users must be registered and logged in to comment should be unchecked. Since only registered members can comment, you lessen the likelihood of spam. If you check this box and allow anyone to register before they submit a comment, you increase the opportunity for automated spammers to leave thousands of links in your blog.

Review each of the remaining statements and determine if you want to allow it or not. By selecting the box next to each state-

NOTES

_____

_____

_____

_____

_____

_____

ment you are agreeing with the action it denotes. A detailed explanation of each comment item can be found in the Word-Press codex at http://codex.wordpress.org/Settings_Discussion_Screen.

You should have a spam preventer plugin installed on your site to limit the amount of spam comments that may appear. Your site has Akismet installed, but you will need to activate it. Other spam prevention plugins can be found in the WordPress plugin repository.

## Media

Media refers to any items uploaded to your site including images, embeds and files like .pdf.

*When you upload an image, it is automatically saved in every size listed unless you have changed the specific file size to 0 by 0.*

Image Sizes

You can determine the default sizes for your images.

Media Settings

Image sizes

The sizes listed below determine the maximum dimensions in pixels to use when adding an image to the Media Library.

Thumbnail size — Width 150 Height 150
☑ Crop thumbnail to exact dimensions (normally thumbnails are proportional)

Medium size — Max Width 0 Max Height 0

Large size — Max Width 0 Max Height 0

NOTES

_____
_____
_____
_____
_____
_____

## Permalinks

WordPress offers you the ability to customize the appearance of your website permalinks. The shaded URL is an example of the information within each permalink will appear.

Setting up your permalinks from the beginning is important for SEO, since your pages will be indexed according to their URL. We go into more detail on permalinks for SEO in Chapter 12. But there are some important details you should note when setting up your permalink structure.

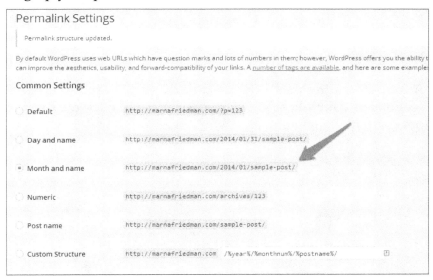

The major permalink items you should be aware of include:

## NOTES

_____

_____

_____

_____

_____

_____

- Leverage your keywords in a descriptive manner when creating titles to make the best use of the three to five words
- Google treats hyphens as spaces between words and underscores as additional parts of words, so use hyphens to separate words instead of underscores

Once the General Settings are complete, you can navigate the other sections of the dashboard.

All of the sections on the Dashboard, except for the menu, can be moved around to customize your view. In addition, by selecting Screen Options (in the top right hand corner of the screen), you can select, and de-select, the items you want displayed on your screen.

And in WordPress 3.8, you can quickly click a link to different sections of your site to get started, or you can click the dismiss option to navigate yourself.

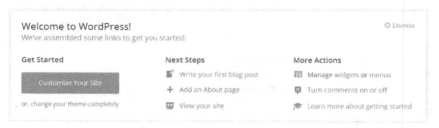

You can click the Help tab, next to Screen Options (see screenshot on next page), to quickly launch help from the WordPress Codex on each screen's topic. The left hand side of the Help screen offers quick links to help on each of the sections dis-

NOTES

_____

_____

_____

_____

_____

_____

played on the screen, while the right side provides hyperlinks to further documentation and support forums. The help options change according to the section of the dashboard that is active.

## Posts

As you continue to navigate the dashboard, you should explore each section to learn what it offers. This is good to do before you install any themes, since they may add even more sections and functionality to your dashboard.

Posts refer to blog posts that are typically displayed in reverse chronological order, with the most recent first. They can also have categories and tags associated with them. You can filter posts by date or category. And by hovering the post title, you will also be presented with options to edit, quick edit, trash or view the post.

While you can edit a post, one of the nice options is Quick Edit, which enables you to quickly modify some of the features without having to open the post. Items you can change include:

NOTES

_____

_____

_____

_____

_____

- title
- slug
- categories
- add tags
- change the date the post will be published
- deselect the ability to allow comments or pings
- password protect or make a post private
- make a post sticky - it stays at the top of your posts

Posts can be scheduled, private or password protected by modifying the options in the Publish section.

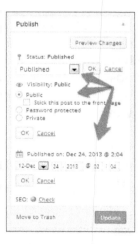

## NOTES

_____

_____

_____

_____

_____

_____

In the post editor, you can use Visual mode or Text mode. The visual editor (see above) will only be shown in Visual mode. The box furthest to the right on the top row is called Kitchen Sink, and when clicked will show the second line of options.

Take some time to navigate each symbol to understand what it does and how it can help you.

You can also see that there are unique icons in the top menu of my editor, which are unique to the theme I am using. Plugins and themes can add more icons to your editor for additional features, such as an IDX to add listings to a page or post.

The Omega icon offers the ability to use special characters in your pages or posts without formatting changes.

NOTES

# Real Estate Detail

When writing posts about real estate, you can categorize your posts with the categories of your business, i.e., location, type of homes, price, etc. Tags can add an extra level of detail.

Category:   broad based topics to organize the content of your posts, similar to a Table of Contents. Create your category list before you start, since your posts will be indexed and changing them later can be costly, time consuming and basically - a nightmare!

Tag:   detailed "keywords" to further organize the content of your posts, similar to an Index

NOTES

_____

_____

_____

_____

_____

_____

_____

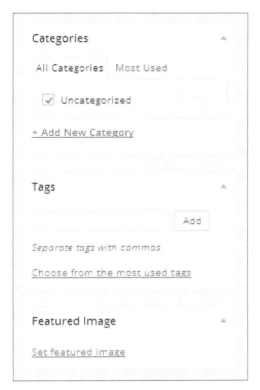

While you may choose keywords for categories and tags, keep in mind that your content should be user focused, and not SEO focused. These terms should relate to the page or post content.

NOTES

_____

_____

_____

_____

_____

_____

# Media

Media is where you upload all items to add to your posts and pages.

Most themes will also offer a "Featured Image" which can be found in the lower right corner of your Post dashboard. If it isn't there, make sure "Featured Image" is enabled in the screen options on the top right. You can upload an image to your post which will then be displayed on your Home Page in a "prefor-matted" size for your theme.

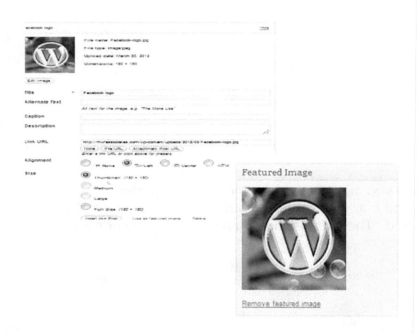

NOTES

_____
_____
_____
_____
_____
_____

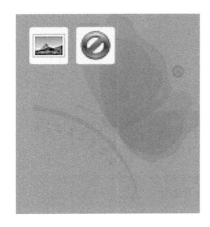

*By clicking on the image, you can then select the symbol on the left over the image to edit the image. Click on Advanced Settings for more options.*

Size                      You can modify the size of the image

Image properties          You can type numbers in the boxes to create an image border and offset the image from text. The data you enter will populate in the Styles box and you can then go back and edit that by changing the border color and style. You can change the appearance by changing the border style and color (see illustration on page 38).

Visit http://w3schools.com/css/css_border.asp for more information and options.

NOTES

_____

_____

_____

_____

_____

_____

# Real Estate Detail

You can link images/logos/pdf files to property listings and other websites by changing the link URL to their website URL.

**N**OTES

Borders are a nice way to style property images and differentiate them from other images used on your website.

## Pages

Pages are for information that is constant, like an "About" page.

Your real estate website should include Pages that are relevant to your business such as Buyers, Sellers, Blog, Contact

NOTES

_____

_____

_____

_____

_____

_____

# Comments

As your readers comment on blog posts, you will see a number next to the Comments section of your Dashboard. You can choose to approve, answer, delete or mark the comment as spam.

# Appearance

The Appearance section offers you the ability to modify the appearance of your website. Themes will be discussed in the next chapter.

| | |
|---|---|
| Widgets | Widgets are a function of the theme and can be changed, deleted or added. |
| Menus | Menus are also a function of the theme and some themes offer multiple menu locations while others offer one. However, you can also create multiple menus as well as custom menus. As you create a menu you can determine if it is the current menu by selecting it in the dropdown on left of your menu Dashboard. |

Theme Options is a function of your selected theme as are Background and Header. These will only appear for specific themes.

| | |
|---|---|
| Editor | You can view your theme files through this option. However, you should not edit any of these files from within WordPress since this can cause your site to "break", and should |

NOTES

_____

_____

_____

_____

_____

_____

only be edited through FTP or File Manager (with the exception of your stylesheet). Always backup your site before you edit any files, and make a copy of the original file before you make any changes.

## Plugins

Plugins add functionality to your site. They will be covered later in this book.

## Users

This is where all of the users of your website can be found. If you have other people helping you with your real estate website, you can assign different roles to them which will limit the access they have. While there are plugins that can expand the roles available on your site, here are the default roles:

| | |
|---|---|
| Administrator | Total access to the site |
| Editor | This person can write posts and can also manage other people's posts |
| Author | Can publish and manage their own posts |
| Contributor | This person can write and manage their posts but cannot publish them |
| Subscriber | Much like a magazine subscription, this user can only manage their profile |

The Dashboard view and access changes with each role.

NOTES

_____

_____

_____

_____

_____

_____

# Tools

You can import and export data through the Tools option.

_____

_____

_____

_____

_____

_____

# WordPress Themes

Learn how to expand the functionality of your WordPress website

- 📖 Understanding Themes
- 📖 Selecting a Theme
- 📖 Installing a Theme
- 📖 Activating a WordPress theme

*Definition*

*The WordPress Theme system is a way to" skin" your weblog. A WordPress Theme is a collection of files that work together to produce a graphical interface with an underlying unifyng design for a weblog.*

*http://wordpress.org*

NOTES

_____

_____

_____

_____

_____

_____

## Selecting the right framework

Themes can provide the look and feel of your website while also offering functionality. WordPress is the platform that the theme works with, it's another component in building your website.

# Understanding themes

Now that you have installed your WordPress website, it will help to understand what a theme is. The default theme, Twenty Fourteen, is automatically installed with WordPress and provides some additional functionality for your website including the ability to change the header, background and include several widgets within your site. You can begin to create a list of functions that you would like included in your website. Some of these may be must haves while others are "nice to haves". These lists will help you in your theme selection. A WordPress developer may be able to customize some functionality for you while a graphic designer can custom design your blog and/or website. Some other options include:

- [ ] **99 Designs**
  **http://99designs.com/**
  99 Designs offers crowdsourced designs for websites, logos and more. You host a design contest, create a design brief and select a package. You select the winning designs.

NOTES

_____

_____

_____

_____

_____

_____

☐ **crowdSPRING**
**http://www.crowdspring.com**
Offers you the ability to tap into a global pool of creatives offering designs for logos, web, product packaging and design, and much more.

If you want to design your own theme, and have it converted to WordPress, you can create a PSD file (Photoshop) and have it converted by a professional service. Some of these services include:

☐ http://xhtml.pixelcrayons.com/
☐ http://www.markupbox.com
☐ https://www.csschopper.com/
☐ http://www.riaxe.com

NOTE: *unlocking a technology tidbit for you*

*Remember to consider your IDX plugin when custom designing a theme*

## Responsive Theme

More than 80% of the world's population now has a mobile phone and 91.4 million of those are smartphones. Research shows that 147.2 million will use tablets by 2015. Given these statistics, you should ensure that your website can be seen and

NOTES

_____

_____

_____

_____

_____

_____

used on multiple platforms. While many WordPress plugins offer the ability to make your theme "mobile friendly", it can effect some of the functionality. But a responsive theme is an approach to web development that enables a website to "respond" to the technology it is being viewed on. A responsive theme quickly adjusts to the new platform and generates a view "fit" for that platform. The view may not be exactly the same across each platform, but provides a better user experience.

A responsive theme is a mobile friendly site, but is not a native app.

# Selecting a Theme

Making a decision on which theme to use can be overwhelming. This is where a mindmap and/or wireframe can help. Creating a checklist of features can also help (see form on page 48). There are thousands of free themes available on the WordPress repository as well as premium themes from several different theme vendors. The biggest difference between most free themes and premium themes is support. In addition to support, some premium themes include the psd files for you to modify. Premium themes usually include:

- Installation & Upgrade and Forum Support
- Built in SEO and Website Optimization
- Built in functionality instead of using a plugin

Some premium theme vendors include:

- ☐ http://bizzthemes.com/
- ☐ http://colorlabsproject.com/
- ☐ http://www.elegantthemes.com/

NOTES

_____

_____

_____

_____

_____

_____

- ☐ http://gorillathemes.com/
- ☐ http://www.inkthemes.com/
- ☐ http://ithemes.com/
- ☐ http://www.pagelines.com
- ☐ http://www.studiopress.com/
- ☐ http://templatic.com/
- ☐ http://themeloom.com/
- ☐ http://www.themeskingdom.com/
- ☐ http://themetrust.com/themes
- ☐ http://themify.me/themes
- ☐ http://www.woothemes.com/

You can also purchase real estate specific themes from:
- ☐ http://agentevolution.com
- ☐ http://realtycandy.com

In addition to premium theme vendor sites, there are theme marketplaces that sell themes from different theme developers. While they typically include support, the support may or may not come from the marketplace. Unlike theme vendor sites, marketplace sites usually include more information like how many times a theme has been sold and user reviews. Some premium theme marketplaces include:
- ☐ http://www.mojo-themes.com/
- ☐ http://themeforest.net/category/wordpress
- ☐ http://wpmarket.net/

NOTES

# Theme Selection

This chart helps determine a theme choice for your website

| Theme Name | URL | Page Template | Custom Header | Background | Widgetized Homepage | Responsive | Shortcodes | Slider | Color Options | Custom Sidebars | Event Custom Post Type | Real Estate Functionality | Property Search | Contact Form | Comments |
|---|---|---|---|---|---|---|---|---|---|---|---|---|---|---|---|
| | | | | | | | | | | | | | | | |
| | | | | | | | | | | | | | | | |
| | | | | | | | | | | | | | | | |
| | | | | | | | | | | | | | | | |
| | | | | | | | | | | | | | | | |
| | | | | | | | | | | | | | | | |
| | | | | | | | | | | | | | | | |
| | | | | | | | | | | | | | | | |
| | | | | | | | | | | | | | | | |

# Installing a Theme

Once you decide on a theme, you will need to install it on your site. Depending on your theme, you can usually install it from the Themes option in the Appearance section of your WordPress Dashboard.

If you are installing the theme from the WordPress repository, select the options you would like from the list, and review the results. Each of the themes will include a description of what is offered which can help you make your selection. To see a larger version of the theme, just click the Preview hyperlink. Once you make your selection, click the Install hyperlink to install it on your website.

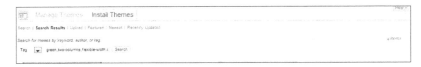

NOTES

_____

_____

_____

_____

_____

_____

_____

Installing a premium theme can be as easy as downloading the theme from the premium vendor's website and then uploading to your website. Sometimes the vendor includes additional files in their download, including the .psd files. Read their documentation for specific instructions on how to install the theme.

Once you have installed your theme, you can begin to customize it through your Appearance Dashboard. If your theme includes a Theme Options Dashboard, it will appear in your Dashboard Menu when you install and activate the theme. If you have purchased any premium themes, you will want to remember your user ID and password to access their Support forums. You should also make sure to keep a record of your receipt numbers for future reference.

You can install multiple themes, but you can only activate one at a time. You can preview a theme on your site after you install it by clicking on the live preview button. Once you activate a theme, you can begin customizing it.

NOTES

If after activating the theme, you decide to change it, you can just upload another one and activate that. Some things to keep in mind when you switch themes:

- any customization you did on the current theme will be gone when you activate the new theme - it will still be in your database, but not visible on the new theme, i.e., menus
- you will need to activate any custom menus you created and assign a location
- you will need to upload logos and favicons
- shortcodes will probably be different, so you will need to update any you created on the previous theme
- page templates will be different, so your content may look different
- you will probably discover other differences, just be sure to remember any customizations you created, so you can update them with a new theme.

NOTES

_____

_____

_____

_____

_____

_____

**NOTES**

_____

_____

_____

_____

_____

_____

_____

# WordPress Plugins

Learn how to expand the functionality of your WordPress website

- 📖 Understanding plugins
- 📖 Selecting plugins
- 📖 Installing plugins
- 📖 Plugins for real estate business functionality

*Definition*

*Plugins are tools to extend the functionality of WordPress. Plugins offer custom functions and features so that each user can tailor the site to their specific needs.*

*- http://wordpress.org*

NOTES

_____

_____

_____

_____

_____

_____

Plugins can quickly expand the functionality of your Word-Press real estate business website. These plugins can enhance the user experience, expedite property search, integrate social media and much more.

# Understanding plugins

Now that you have installed your WordPress website, it will help to understand what a plugin is. WordPress is the platform you will use, but much like a theme, plugins are another building block. While your theme expands the design of your site, plugins expand the functionality. As we discussed in Chapter 4, some of this functionality may be built into your premium theme, while plugins may offer this where it doesn't already exist.

There are specific functions that should be included in any site, but there are several plugins that offer each of these functions. You need to make an informed decision on these.

# Selecting Plugins

How do you determine the best plugin for your site? Like the chart on page 48, you should create a chart for each of the plugin categories that can help you make an informed decision. As you navigate the different plugins available in each category, you will be able to identify the features that are important to you and your readers. While you may want certain features, are they important to your visitors, or important to you? The answer to this question should always be your visitors.

NOTES

_____

_____

_____

_____

_____

_____

Key plugin functions for any site are:
- SPAM preventer
- SEO
- Backup
- Analytics
- Contact form
- Page navigation
- Cache
- Editorial Calendar

# How do you decide which plugin?

The WordPress plugin repository provides an opportunity for a lot of information that can help you make your decision. But you need to understand how to use this information.

- Does the plugin offer the functionality you need?
- Is the plugin compatible with the current version of WP?
- How many times has it been downloaded?
- When was it last updated?
- What version of WordPress does it require?
- What version of WordPress is it compatible to?
- Is the developer supporting the plugin?
- What are the forum comments?
- Does it conflict with other plugins you are using?
- Are there screenshots?
- Is there documentation?

NOTES

_____

_____

_____

_____

_____

_____

Another consideration would be if the plugin is a free version of a premium plugin. Typically this is indicative of a developer that is supporting the plugin, at least the premium plugin. You can test the free version and upgrade to the premium when you are ready.

# Installing Plugins

Once you have selected a plugin, you can click on the Install-Now button. Since this installation can affect your website, WordPress asks you to confirm this decision. Once you complete the installation, you will still need to activate the plugin.

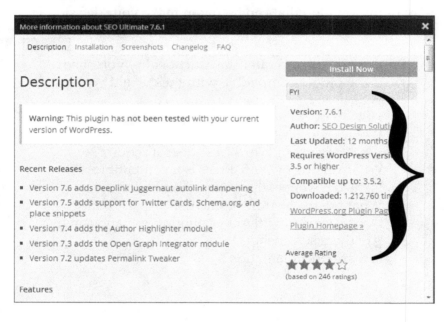

NOTES

**Plugin musts:**

- You should be sure you have access to FTP or your host service whenever you install a plugin, in case it "breaks" your site and you need to uninstall it.

- Always deactivate a plugin before deleting it. And check to see if the developer has added any other details you need to be aware of if you delete the plugin.

- Plugins can conflict with other plugins and themes. Keep track of when and which plugins you have installed. If you do have a conflict, you typically have to deactivate all of your plugins and then reactivate one by one to identify the conflict.

# Basic Plugins for Every Website

Earlier in this chapter, we discussed key plugins that every website should have. Now we will suggest different plugins from the WordPress repository for each of those functions.

## SPAM PREVENTER

WordPress comes with Akismet preinstalled, and this is a great SPAM preventer. You will need to have a key to activate this plugin.

Some other SPAM preventer plugins include:
- ☐ Anti-SPAM by Cleantalk
- ☐ SPAM Free WordPress

NOTES

_____

_____

_____

_____

_____

_____

# SEO

Search Engine Optimization (SEO) is an important aspect of every website and requires much more than a plugin. But a plugin can help you set your site up properly for SEO.

- [ ] All in One SEO Pack
- [ ] Greg's High Performance SEO
- [ ] WordPress SEO by Yoast

## ANALYTICS

Analyzing how visitors are using your website can help you improve their experience and increase your impact. You can do this with Google Analytics, and we will review how to set this up later in this book, but you can also use plugins on your website to help you with this including adding your analytics to your WordPress dashboard.

- [ ] Clicky Analytics
- [ ] Google Analytics Dashboard for WP
- [ ] Google Analyticator

## BACKUP

We will go into detail about backups in a later chapter, but in the meantime, here are some backup plugins that are available on the WordPress repository for you to use.

- [ ] UpdraftPlus - WordPress Backup and Restoration
- [ ] WP-DB-Backup

NOTES

_____

_____

_____

_____

_____

## CONTACT FORM

A contact form is a great way to capture emails and contact information for future follow-up. This is one of the areas where I recommend a premium plugin, but there are several free contact form plugins available also.

- ☐ Contact Form
- ☐ Contact Form 7
- ☐ Contact Form Maker

## POLICY

You should include a Privacy Policy and Terms of Use on your website. You can include these as pages on your website and link to them in your footer, or use a plugin. Since these documents pertain to legal issues, it is **advisable to consult a professional**.

- ☐ Auto Terms of Service and Privacy Policy
- ☐ Agreement

## CACHE

There are lots of different things that can slow your website down. A cache plugin can help you improve the speed of your website, which in turn can help the user experience and SEO results.

- ☐ Quick Cache
- ☐ W3 Total Cache
- ☐ WP Super Cache

## NOTES

_____

_____

_____

_____

_____

_____

## SOCIAL MEDIA INTEGRATION

There are several social media integration methods, and using a plugin is one of them. Here are some plugins that will help you integrate your social media on your website.

### FACEBOOK
- [ ] Add Link to Facebook
- [ ] Flexo Facebook Manager
- [ ] Simple Facebook Connect

### GOOGLE PLUS
- [ ] Google Plus Google
- [ ] Google Plug Widget

### LINKEDIN
- [ ] LinkedIn Share Button

### PINTEREST
- [ ] Pinterest Badge
- [ ] Pinterest Pin It Button
- [ ] Pretty Pinterest Pins

### TWITTER
- [ ] Twitter Badge Widget
- [ ] Tweet Blender
- [ ] Xhanch MY Twitter

NOTES

_____

_____

_____

_____

_____

_____

Some plugins include additional settings and configuration in order to use them. If so, the plugin will note that with a Settings option. You should review the plugin documentation before moving forward to ensure that you are installing the plugin correctly.

Other plugins require additional resources in order to function properly, which may incur an additional cost. This information is usually in the plugin documentation, but if you did not review that before installing, you may not be aware until you try to activate the settings and complete the installation.

NOTES

_____

_____

_____

_____

_____

_____

# oEMBED

While there are several YouTube plugins available, you should also understand that WordPress offers a feature called oEmbed. This feature enables you to paste a URL from specific sites into your WordPress site and it will automatically paste a video or image onto your post or page. You do not need a plugin to use the oEmbed function.

NOTE: *unlocking a technology tidbit for you*

*For security purposes, you can only embed URLs matching an internal whitelist. So if a URL is "private", you will not be able to embed it.*

How is this done?
1. Go to the application (see list below)
2. Find the image/video you want
3. Copy the URL from the browser
4. Go back to WordPress page/post
5. Ensure you are in HTML mode
6. Paste URL
7. Save

NOTES

_____

_____

_____

_____

_____

_____

Websites that offer the oEmbed functionality include:

- blip.tv
- DailyMotion
- Flickr (both videos and images)
- FunnyOrDie.com
- Hulu
- Instagram
- Photobucket
- PollDaddy
- Qik
- Revision3
- Screenr
- Scribd
- SlideShare
- SmugMug
- SoundCloud
- Twitter
- Viddler
- Vimeo
- WordPress.tv (only VideoPress-type videos for the time being)
- YouTube (only public videos and playlists - "unlisted" and "private" videos will not embed)

NOTES

_____

_____

_____

_____

_____

_____

# The oEmbed Process

1. Type your content/post and then in text mode
2. Select the item you want to include from one of the websites listed on page 63.
3. Copy the URL

4. Paste the URL into your page or post - make sure you are in text mode

5. That's it. You can preview to see if it worked correctly. If it didn't, confirm that you pasted the URL onto your website while you were in text (HTML) mode and not visual.

NOTES

# WordPress IDX Plugins

Learn how to expand the functionality of your WordPress website to include MLS listings on your website

- 📖 Understanding IDX
- 📖 Selecting an IDX
- 📖 Installing plugins
- 📖 Using IDX

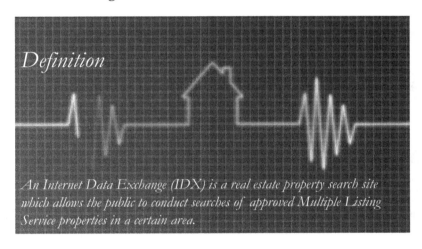

*Definition*

*An Internet Data Exchange (IDX) is a real estate property search site which allows the public to conduct searches of approved Multiple Listing Service properties in a certain area.*

NOTES

_____

_____

_____

_____

_____

_____

# Understanding IDX

An IDX plugin will enable you to integrate MLS listings onto your website. Your MLS regulates and approves the IDX providers that members can use. Basically IDX is a data feed, and until recently, was the only way of accessing MLS data for websites. The IDX feed is typically updated nightly through FTP (file transfer protocol) downloads. This process can take a significant amount of time, and can sometimes cause delays in listing updates.

You may also hear the term RETS being used. RETS (Real Estate Transaction Standard) is similar to cache on your own website. The database doesn't need to be downloaded to your site daily, only the information being requested/queried. This lowers the amount of bandwidth your site uses, and as a result, speeds it up. Because RETS is newer technology, it is not available everywhere and not every IDX plugin offers a RETS option. For those that do, it is usually more expensive than the IDX.

As the technology evolves, RETS will probably become more of an industry standard and therefore, more readily available.

# Selecting an IDX Plugin

So now that you understand the differences between IDX and RETS, you might want to keep it *on file* for future reference. The selection of an IDX is the most important one you will make for your website. Once you install and activate the plugin, depending on how it is set up, it can affect your SEO and indexed

NOTES

_____

_____

_____

_____

_____

_____

pages. If your listings are currently on a subdomain, be sure to make note of that, since you will want your new IDX to use the same set-up. Some things to consider when selecting an IDX plugin:

- 📖 Is it approved by your MLS
- 📖 What is the support
- 📖 How easy is it to integrate into a theme
- 📖 What are users saying about it - don't just take the vendor's word for it
- 📖 Cost
- 📖 Is it responsive? Known issues - speed, SEO, etc.
- 📖 Are there *ready made themes* to make launching a site easier
- 📖 Plugin conflicts and can you use add other plugins to your site, i.e., events, shortcodes, etc.
- 📖 Specific requirements?

I am sure you may have more questions than those above, but the above are fairly global. My recommendation would be to create a spreadsheet of your questions and list each plugin you are considering with how they answer your needs checked off, similar to the one on page 74. This should help you make a more informed decision.

*Once you select a plugin and begin to use it, you will be creating several property links to community pages and more. If you decide to switch, all of those links would need to be changed.*

NOTES

_____

_____

_____

_____

_____

_____

Not all IDX plugins will work with WordPress, and those that do offer a variety of features. Here is a list of some of them, and some of their features.

**Displet 2.0**
**http://displet.com/idxrets-search/**
- Responsive
- API integrations
- Indexable Listings
- Multiple Widgets including Quick Search
- Round Robin Lead Assignment
- User Saved Searches
- Customizable CSS
- Zapier Automated Tasks

**dsIDXpress**
http://www.diversesolutions.com/
- Multiple widgets including Quick Search
- Indexable Listings
- Lead Capture
- Open House data*
- Shortsales and REO
- School data
- SEO optimization*
- Customizable CSS
- Mobile*

*only certain programs offer these features

NOTES

_____

_____

_____

_____

_____

**FlexMLS**
https://www.flexmls.com/
- SEO Optimized
- Customizable Search Widget
- Multiple Widgets including Quick Search
- Unlimited Landing Pages

**IDX Broker/Platinum**
http://www.idxbroker.com/
- myAgent mobile app*
- Responsive*
- SEO tools
- Lead Capture
- Lead Routing for Teams/Offices
- Visitor Statistics
- Customizable Templates
- Customizable Search
- Indexable Listings
- Multiple Widgets including Quick Search
- Customizable Widgets
- IDX Mapping
- User Saved Searches
- Multiple MLS available*

   *myAgent mobile app on IDX Broker
   *IDX Platinum responsive on responsive themes
   *Multiple MLS available at additional cost

NOTES

_____

_____

_____

_____

_____

_____

## iHomefinder

http://www.ihomefinder.com/product/optima-express/

- Lead Capture
- Quick Search Widget
- User Saved Searches
- Open House data
- Mobile website
- Facebook Search
- Interactive Map Search*
- VIP Support*

*these options only available on certain programs

## Placester

https://placester.com/

- Responsive
- SEO Optimized
- Customizable Templates
- Indexable Listings
- RETS integration
- Unlimited Landing Pages
- 

## Showcase IDX

http://showcaseidx.com

- Responsive
- Live Map Search
- Lead Capture
- Social Media Sharing
- Combined MLS Feeds

NOTES

_____

_____

_____

_____

_____

_____

- Foreclosures and Short Sales
- Saved Searches
- SEO Optimized
- MLS Compliant
- Indexable Pages
- Unlimited Landing Pages
- Multiple Widgets including Quick Search

## SpotOnConnect
http://spotonconnect.com/
- Responsive
- Lead Capture
- CRM Integration
- Indexable Listings
- Saved Searches
- SEO Optimized
- Unlimited Landing Pages
- Several search queries to use on landing pages
- Multiple Widgets including Quick Search

## WolfNet
http://wolfnet.com/web-plugins/
- SEO Optimized
- Indexable Listings
- Customizable Search
- Quick Search Widget

Each of these IDX plugins offers additional features and benefits that continue to change as they upgrade and respond to feature requests.

NOTES

Real Estate Detail

*While there are several WordPress plugins available to add MLS listings to your website, you will still need to pay the applicable MLS Fees which are payable directly to the MLS*

# Installing the IDX Plugin

Once you have selected a plugin, you will need to follow the directions on how to use it. Some of them are easier to implement than others, which should be considered before you make a final selection on IDX and theme. If you need help in integrating the plugin on your WordPress website, you may find that the theme, as well as your host provider may incur additional costs.

Here are some tips on the installation process for a few of the plugins:

**IDX Broker/Platinum**
This plugin requires a more detailed installation process.
1.  Creating a wrapper so that your IDX pages look like the rest of your WordPress website
2.  Setting up a custom subdomain with a CNAME record
3.  Creating custom search lists for your widgets (optional)
4.  The remainder of your setup is similar to other IDX plugins with creating neighborhood pages, saved links, etc. as well

NOTES

_____

_____

_____

_____

_____

as SEO set-up.

**ShowcaseIDX**

1. This plugin activation is similar to most others in that you upload the plugin, activate it, add your API key from the vendor.
2. SEO set-up for search
3. Create "hotsheets" for neighborhood pages, landing pages, etc.

**SpotOn Connect**

1. This plugin activation is similar to most others in that you upload the plugin, activate it, add your API key from the vendor.
2. SEO set-up for search
3. Create "hotsheets" for neighborhood pages, landing pages, etc.

NOTES

_____

_____

_____

_____

_____

_____

# WordPress IDX Plugins

This chart compares plugins that expand the functionality of your site.*

| Plugin Name | URL | Approved by my MLS | Responsive | Built-in Lead Capture | Widgets | Shortcodes | SEO Optimized | Map Search | Analytics | Agency/Office Functionality | CRM Integration | Round Robin | Support Available | Themes Available | Easy to Integrate | | | Cost | Comments |
|---|---|---|---|---|---|---|---|---|---|---|---|---|---|---|---|---|---|---|---|
| Displet 2.0 | http://dsplet.com/idxrets-search | | | | | | | | | | | | | | | | | | |
| dsIDXpress | http://www.diversesolutions.com | | | | | | | | | | | | | | | | | | |
| FlexMLS | https://www.flexmls.com | | | | | | | | | | | | | | | | | | |
| IDX Broker | http://www.idxbroker.com | | | | | | | | | | | | | | | | | | |
| IDX Broker Platinum | http://www.idxbroker.com | | | | | | | | | | | | | | | | | | |
| iHomefinder | http://www.ihomefinder.com | | | | | | | | | | | | | | | | | | |
| Placester | https://placester.com | | | | | | | | | | | | | | | | | | |
| Showcase IDX | http://showcaseidx.com | | | | | | | | | | | | | | | | | | |
| SpotOnConnect | http://spotonconnect.com | | | | | | | | | | | | | | | | | | |
| | | | | | | | | | | | | | | | | | | | |
| | | | | | | | | | | | | | | | | | | | |
| | | | | | | | | | | | | | | | | | | | |

Chart compiled by Marnie Friedman · @marnafriedman · http://impressivesolutions.com

# Plugins for Real Estate

Learn how to expand the functionality of your WordPress website
    📖 Types of plugins for real estate
    📖 Selecting plugins

We've seen plugins that can expand the functionality of a basic WordPress website as well as how IDX plugins can quickly convert a basic WordPress website into a real estate business website. Now we will look at more plugins that can be added to your website to provide more information and engagement for your visitors.

# Types of plugins for real estate

There are so many plugins available that it is easy to overload your site with them which can result in a slower site, possible code conflicts between plugins, and ultimately confusion for your visitors. So review the information and make your selections wisely.

NOTES

_____

_____

_____

_____

_____

_____

Here are some items you may want to include in your website:

## Property Listings

Sometimes you may just want a website for your own listings, rental or commercial properties. There are plugins that can be added to your WordPress website that will offer this functionality.

☐ FireStorm Real Estate Plugin
☐ Real Estate Agency for WordPress
☐ Real Estate Website Builder
☐ WP Property

## Financial Calculators

You can quickly add mortgage calculators to your website so that your visitors can see what a potential home might cost them. Other calculators include affordability, closing costs, and more.

☐ Advanced Real Estate Mortgage Calculator
☐ Calculator Pros Calculators
☐ CT Mortgage Calculator

## Maps

Maps should be a given on a Real Estate website. If your IDX plugin does not include maps, then you might want to add a map plugin.

☐ Comprehensive Google Maps
☐ Google Maps GPX Viewer

## Notes

_____

_____

_____

_____

_____

_____

## TESTIMONIALS/REVIEWS
Real Estate is a referral business, and testimonials/reviews can certainly help with referrals. Having this functionality on your website can help you follow up on this task.
- Rich Reviews
- WP Customer Reviews
- WP Reviews

## DIRECTORY
Your buyers and sellers frequently look to you for information on local vendors including mortgage companies, home inspections, repair, plumbers and more. As a convenience to their customers, many real estate agents offer a list of these, and an online directory would probably be easier to maintain.
- Connections
- Business Directory
- Directory & Classified

## EVENTS
As a local expert, you may want to add neighborhood events to your website.
- All in One Event Calendar
- The Events Calendar
- WordPress Event Calendar

## NOTES

_____

_____

_____

_____

_____

_____

## Editorial Calendar

Blogging has become a very hot topic for real estate agents. Sharing information with your community can increase your reputation, and the regular addition of content to your website can help your search engine results. An editorial calendar can help you manage this process, especially if you have someone helping you.

- ☐ Betaout Content Cloud
- ☐ Edit Flow
- ☐ Editorial Calendar

# Selecting Plugins

Review each of the categories to determine which functionality you would like to add to your website. Prioritize the plugins so that as you install them, if there is a conflict, you will know which ones you really want to have. Another factor to consider is how easy the plugin is to use and integrate with your theme. If the plugin requires a significant amount of customization, then you may want to reconsider it.

Other factors to consider:
- Do you really need it?
- Is this functionality available in your theme?
- Does the plugin have a premium option if you choose to upgrade?
- Will you use it?
- Is the plugin being supported?
- Does my state license allow me to share this information on my website?

## Notes

_____

_____

_____

_____

_____

# Page Builder

Learn how to create page layouts for your WordPress website
- 📖 Understanding Page Templates
- 📖 Using Page Builder Plugins

## Understanding Page Templates

WordPress offers page templates that can change the "appearance" or layout of different pages on your website. The default page template in WordPress is usually page.php. Your theme may also include other custom page templates, such as pages for:

- Home
- Team
- Portfolio
- Blog
- Contact

As you explore different themes for your website, determine what pages you might need and if those page templates are included. If not, you would have to create them, or possibly use shortcodes. But there is another option, and that is using a page builder.

NOTES

_____

_____

_____

_____

_____

Page Builders have been around for a while, but have become increasingly popular as WordPress users seek to expand the functionality of their website.

If you choose to create your own page templates, refer to the WordPress codex at http://codex.wordpress.org/Page_Templates.

# Using Page Builder Plugins

Several premium themes now include page builder options, but if they do not, there are free page builder plugins available on the WordPress repository as well as premium plugins. The plugins on the WordPress repository include:

- Aqua Page Builder
- IG PageBuilder
- MotoPress
- Page Builder by SiteOrigin

Page Builder plugins can be activated on any page and once activated, you use them instead of the WordPress editor. They enable you to quickly create unique page layouts, and once created, can often be saved as "templates" to reuse the same layout on similar pages, i.e., services.

Understanding the pros and cons are important as you begin your website development.

NOTES

_____

_____

_____

_____

_____

**Pros**

quick page layouts
templates without code
see what's possible

**Cons**

too much can be overwhelming
may not be maintained
layouts may differ in themes

Look at the features offered in the different page builders to see if the functionality you are looking for is included. If not, move on to another one until you find the one that includes what you are looking for. You can always have custom page templates created based on the pages you create using a page builder, as long as the functionality is included in your theme or you have your website custom coded, i.e., toggles, tabs, etc.

Real Estate Detail

*You can create page templates for subdivision landing pages that offer consistency and quick formatting on your website.*

Most of the page builder plugins are drag and drop offering several options. Some of them may conflict with your theme and other plugins, so you should test them to ensure that all of the functions are working.

NOTES

_____

_____

_____

_____

_____

# Premium Plugins

As we discussed, there are also premium page builder plugins that you can purchase to use on your site. Since these are premium plugins, they typically include support, which might be a key factor in your decision.

☐ **Visual Composer by WP Bakery**
http://vc.wpbakery.com/

☐ **Elegant Builder by Elegant Themes**
http://www.elegantthemes.com/

☐ **Builder**
http://themify.me/builder

☐ **Dynamik Genesis Extender**
http://cobaltapps.com/downloads/genesis-extender-plugin/

☐ **Ether Content Builder WordPress Plugin**
http://ether-wp.com/ether-builder/

NOTES

_____

_____

_____

_____

_____

# CTA & Lead Generation

Learn how to integrate effective calls to action (CTA) and Lead Generation in your WordPress website

- 📖 Understanding CTA
- 📖 Integrating CTA
- 📖 Lead Generation Tools
- 📖 Lead Generation Forms
- 📖 Marketing Automation

# Understanding CTA

*Definition*

*A call to action is a request/direction to 'do something' — often the next step that a website visitor should take toward the purchase of a product or service, contact, download or other action.*

## NOTES

_____

_____

_____

_____

_____

Most real estate professionals have a website, and many of them are not using CTA (calls to action) in a manner that will effect a response. In fact, small businesses in general do not understand how to use a CTA. Some interesting statistics from a recent study by Online Marketing Coach (http://www.onlinemarketingcoach.com/2013-small-business-b2b-study/) are staggering:

- 70% don't display clear calls to action for anything on their home pages.
- 72% don't have any calls to action on their interior pages
- 82% don't reference their social media profiles
- 27% don't include a phone number on their home pages
- 31% of web sites with a phone number don't display it in a prominent place
- 68% don't include an e-mail address on their home pages
- 38% of web sites with an e-mail address bury it on the home page (often in the footer)

Real estate websites CTA include:
- search for homes
- what's your home worth
- market conditions
- local information
- calculators

Seems easy enough? Yet, many real estate websites do not include these. A pretty website is great, but without a CTA, how is a visitor supposed to react? Understanding how to integrate a CTA without overpowering your website is your next step.

## NOTES

_____

_____

_____

_____

_____

# Integrating Calls To Action

Some real estate themes, like those from Agent Evolution, include calls to action as part of their framework (*see example at bottom of page*). And several other themes also include these.

## NOTES

_____

_____

_____

_____

_____

Calls to Action on a real estate website can be integrated in several different locations in order to convert visitors into leads. As you can see from the examples on page 85, there are several options. Now we'll talk about types and locations of these calls to action:

## Phone number

Making yourself available and letting visitors know that, is important. Don't hide your phone number in small print. You can place your phone nunber prominently in the header *and* footer of your website, but you can also place a callout in the center of your homepage letting people know how to get in touch with you. Remember that once you place this phone number prominently on your website, you need to be prepared to answer it.

**Looking to sell your home? Call us at (212) 555-1212**

## Search

You are creating a real estate website. Searching for properties is probably one of the reasons visitors will be coming to your site. If you do not have search functionality on your home page as in example 2 on page 85, then offer a prominent search button with clear instructions as in example 1.

NOTE: *unlocking a technology tidbit for you*

*Always test buttons and links before you publish them to ensure that they "land" where you want them.*

NOTES

_____

_____

_____

_____

_____

And if you are using buttons, consider testing them with A/B Testing to determine what size and color creates conversion. You can use plugins to test this and then track results before you make a final decision.

Factors to consider when using buttons for CTA:
1. Color
2. Text
3. Size
4. Placement

### A/B Testing Plugins
- ☐ Marketing Optimizer for WordPress
- ☐ Nelio AB Testing
- ☐ WordPress Calls to Action
- ☐ WP Experiments Free

## IDX Lead Capture
This is a personal preference. Some real estate agents like for visitors to sign up whenever they search for properties on their website. Others don't like forced registration at all. While others request visitors to register after a specific number of searches. Whatever your preference, be sure to capture the information you need for follow-up.

## Repetitive Information
There is some information that you want accessible throughout your website. And you can use your footer for these CTA's. This can include:

NOTES
_____

_____

_____

_____

_____

- Resources for buyers and sellers
- IDX area/community/subdivision links
- Testimonials
- Social media links
- Contact information

# Lead Generation Tools

In addition to CTA, you can use a variety of lead generation tools on your website. These include:
- forms
- landing pages
- email sign-up
- social sharing
- thank you pages
- Google Analytic conversion goals

Probably one of the most important tools you should use for lead generation is a CRM. By integrating your CRM with your website forms, you can easily have leads populate in your CRM to create an automated system.

## Lead Capture Plugins:
☐ Easy Lead capture Vtiger Webforms
CRM: https://www.vtiger.com

☐ WordPress Leads
CRM: https://www.zoho.com/crm/
CRM: http://www.sugarcrm.com/

NOTES

_____

_____

_____

_____

_____

☐ WordPress-to-lead for Salesforce CRM
    CRM: http://www.salesforce.com/

☐ WP Zoho CRM
    CRM: https://www.zoho.com/crm/

# Landing Pages

Landing Pages can be used for lead capture. One common use for real estate agents is home valuation. You can create a landing page offering a free home valuation which enables you to capture leads, and by integrating the lead capture information from the form into your CRM, you have created an automated task for your real estate business.

As visitors sign up for a home valuation, their data is populated into your CRM for further follow-up. You can also offer ebooks on home staging, home buying tips, etc. and use these for lead capture where an email is required for the download and you then have the email for follow-up. Once the visitor completes their email, there is usually a second form to complete with additional information that also populates to your CRM providing more de-

NOTES

_____

_____

_____

_____

_____

tails you may need for follow-up. The form can request whatever information you want, and depending on your CRM, you can designate which fields from the form should populate.

### Landing Page Plugins
- [ ] Easy Landing Pages
- [ ] Ultimate Landing Page and Coming Soon Page
- [ ] WordPress Landing Pages
- [ ] WP Lead Plus Free Squeeze Page Creator

### Premium Landing Page Plugins
- [ ] ConvertPress
  http://convertpress.com/
- [ ] Launchpad
  http://www.obox-design.com/launchpad.cfm
- [ ] Max Inbound
  http://maxinbound.com/

# Popups

Another tool you can add to your website for lead capture is pop-ups. These are items that can float over your screen to capture the visitor's attention, or are placed on the perimeter of your website and popup or out to capture your visitor's attention and get them to sign up.

### Popup Plugins
- [ ] Arty Popup
- [ ] ITRO Popup Plugin
- [ ] Optin Revolution: WordPress Popup

NOTES

_____

_____

_____

_____

_____

**Side Offer**
- ☐ WP Popup Plugin

**Premium Popup Plugins**
- ☐ OptinMonster
  http://optinmonster.com/
- ☐ Pippity
  http://pippity.com/

# Lead Generation Forms

Now that you see the different types of lead generation tools, you will need to create a form to capture the lead information. Determine the information you want to capture first so that you make sure to include those fields in your form. Think about what you might need:

- Name
- Email
- Phone number
- Address

You can create multiple forms and multiple landing pages to track different campaigns, but all of the information will still be populated into your CRM.

Your IDX solution may provide lead capture forms that you can use to populate into their CRM. You can also check to see if you can modify the form to feed into your CRM, if it is different from the one they offer.

NOTES

_____

_____

_____

_____

_____

There are several form plugins available for WordPress, which we have listed below. You will need to review the documentation to determine if it will work with your CRM.

## Form Plugins
- ☐ Contact Form by ContactUs
- ☐ Contact Form Maker
- ☐ Form Maker
- ☐ FormGet Contact Form
- ☐ Ninja Forms

## Premium Form Plugins
- ☐ Gravity Forms
  http://www.gravityforms.com/

- ☐ Wufoo Forms
  http://www.wufoo.com/

Gravity Forms is a fairly popular form used by several Word-Press developers in their premium themes. As a result, there is a significant amount of information available on how different developers have integrated Gravity Forms to some popular CRM platforms. In addition, Gravity Forms offers add-ons that also help you integrate their plugin with other programs including email, payment gateways, and more. The lead generation forms on page 93-94 is a Gravity Form that integrates with Zoho CRM and MailChimp.

## NOTES

_____

_____

_____

_____

_____

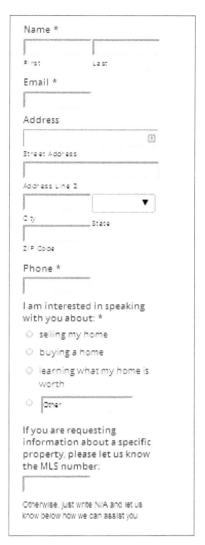

Name *

First        Last

Email *

Address

Street Address

Address Line 2

City        State

ZIP Code

Phone *

I am interested in speaking
with you about: *

○ selling my home

○ buying a home

○ learning what my home is
worth

○ Other

If you are requesting
information about a specific
property, please let us know
the MLS number:

Otherwise, just write N/A and let us
know below how we can assist you.

MailChimp

Settings

MailChimp

Uninstall

**MailChimp Account Information**

MailChimp makes it easy to send email newsletters to your cus
to your MailChimp subscriber list. If you don't have a MailChimp

MailChimp API Key

Save Settings

**Uninstall MailChimp Add-On**

Warning! This operation deletes ALL MailChimp Feeds.   Uni

## MailChimp Feeds   Add New

Bulk action ▾   Apply

☐    Form

☐ ✓   What Is Your Home Worth?

☐ ✓   Information

☐ ✓   Contact

☐    Form

MailChimp Feed

MailChimp list        Marna Sells Real Estate ▾

Gravity Form          Information ▾

Map Fields            List Fields        Form Fields
                      Email Address *    Email
                      First Name         Name (First)
                      Last Name          Name (Last)

Opt-In Condition      ☐ Enable

Options               ✓ Double Opt-In
                      ✓ Send Welcome Email

Update   Cancel

# NOTES

_____

_____

_____

_____

_____

## HOME VALUATION

Complete this information for a free home valuation.

Name*

| First | Last |

Email*

Phone*

## Home Valuation

Address of Property*

Street Address

Address Line 2

City                                State

ZIP Code

Property Type*

Single Family Resident

Condition
- Excellent
- Good
- Fair
- Needs work
- Poor

Bedrooms                      Bathrooms

1                                    1

Additional Rooms/Features

Please list additional rooms and describe any special features and recent upgrades. For example: new roof, new carpet, custom kitchen cabinets, etc

Approximate Size of Home        Approximate Age of Home

square feet

Submit

By submitting this form with your telephone number you are consenting for Mema Friedman and authorized representatives to contact you even if your name is on the Federal "Do-not-call" List.

# NOTES

_____

_____

_____

_____

_____

# Marketing Automation

Many of the CTA and Lead Generation items discussed in this chapter can be automated with programs like InfusionSoft. And there are other marketing automation options available that offer similar "package" programs.

These programs help you create your items that may include landing pages, forms, email newsletters, autoresponders, CRM integration and lead capture through one platform that integrates with your WordPress website.

### Marketing Automation Plugins
- ☐ Jumplead*
- ☐ WP HIVE

### Premium Marketing Automation Plugins
- ☐ **Orbtr**
  http://orbtr.net/

And finally, there are premium Landing Page themes that can help you with your marketing automation and marketing site creation.

- ☐ **OptimizePress**
  http://www.optimizepress.com/

- ☐ **Premise**
  https://purchase.getpremise.com/

## NOTES

_____

_____

_____

_____

_____

## NOTES

_____

_____

_____

_____

_____

# Widgets

Learn how to expand the functionality of your WordPress website
- Types of widgets for real estate
- Selecting widgets

Sometimes you don't need to use a plugin on your website, but you can use a widget instead.

## Types of widgets for real estate

There are so many widgets available for you to use on your website, and there are pros and cons to doing this. While adding HTML code to a text widget can be easier than uploading a plugin that might conflict with your current site, it might be even more of a risk.

While the syndicated real estate battle continues, your using a widget of theirs on your site adds to their ability to maintain their search rankings. The widget text usually has a link back to the syndicated site, and by using the widget, you are agreeing that it is OK for them to benefit from the exposure on your site.

NOTES

_____

_____

_____

_____

_____

A backlink lets Google know that you've used something from another site, and they then offer domain value to that site for providing something of value. The more backlinks a site has, the better it looks to Google, and therefore the higher it ranks in search results.

Keep in mind that according to the terms of use on many of these widgets, you cannot modify them. And while it is great that the syndicated sites make these widgets available for free, they are actually causing you to incur a cost, which is the cost of lower SERP.

So why would I even include this information? Because you need to know, and make informed decisions about what to include on your website. Many of the plugins are also offering the same information, also from syndicated sites. A first defense would be to see if your IDX plugin offers any of this information. If not, you can search the web for alternatives. While the widget you choose may also provide backlinks, it may not be for a direct competitor, and so may not impact your search results.

# Creating a Widget

1. Select the widget you would like to use.
2. Go to the website and create the widget according to their directions.
3. Copy the HTML code that they give you.
4. Go back to your WordPress website Dashboard > Appearance > Widgets.

NOTES

_____

_____

_____

_____

_____

5. Select a text widget and drag it to the sidebar or footer location you would like the widget.
6. Paste the HTML code in the widget, add a title and save.
7. If you want to paste the widget on a page instead, then:
8. Go to the page where you want the code.
9. Select Text option in your editor.
10. Paste the code.
11. Save the page.

# Selecting Widgets

Now that we have explained the pros and cons of using widgets from the larger syndicated real estate websites, here are some resources for you to find widgets that you might want to use on your website.

## Financial Calculators

☐ **FHA Loan Limits**
*http://www.fha.com/fha_loan_limits_widgetcode*
This government agency offers a variety of widgets relating to their loan programs that you can include on your website.

☐ **MortgageLoan**
*http://www.mortgageloan.com/widgets*
Mortgage Loan offers a variety of financial calculators that you can add to your website including mortgage calculator, credit card payoff, how much can you afford and more.

NOTES

_____

_____

_____

_____

_____

_____

☐ **EPPRAISAL**
*http://www2.eppraisal.com/widgets/index.mvc*
Offering a variety of widgets on home values.

☐ **CALCMOOLATOR**
*http://www.calcmoolator.com*
Offering a variety of calculator widgets for financial calculations incuding debt, mortgage, rent vs. buy and more.

## Neighborhood Information

☐ **AREA VIBES**
*http://www.areavibes.com/widgets*
This website offers a variety of widgets focusing on different aspects of your neighborhood including city data and demographics.

☐ **GREAT SCHOOLS**
*http://www.greatschools.org/about/linkToUs.page*
This website offers information on local schools, but keep in mind that even their URL states "linktous".

☐ **WALKSCORE**
*http://www.walkscore.com/professional/badges.php*
This widget is help for downtown and city locations offering an overview of the area's convenience based on a walking score to neighborhood amenities.

NOTES

_____

_____

_____

_____

_____

## Moving Expenses

☐ **MOVING.COM**

*http://www.moving.com/widgets/packing-calculator.asp*
This widget can be used to offer help with sellers on moving costs and packing calculators.

NOTES

_____

_____

_____

_____

_____

_____

NOTES

_____

_____

_____

_____

_____

_____

# Live Chat

Learn how to expand the functionality of your WordPress website

- 📖 Pros and Cons of Live Chat
- 📖 How does Live Chat Work?
- 📖 Live Chat Plugins

We have become a society of instant response. And if you don't answer the phone, a client may move on to another real estate agent. Through Live Chat functionality, you can quickly turn a website visitor into a client.

## Pros and Cons of Live Chat

Live Chat has been used by service businesses for quite a while to instantly respond to questions and yes, complaints. It has often resulted in sales and happier clients. So why not use it for real estate? And yet again we look at the pros and cons of incorporating this functionality.

NOTES

_____

_____

_____

_____

_____

_____

**Pros**
Quick response to visitor inquiries
Lead capture

**Cons**
Cost
Time and staffing

Before you integrate Live Chat on your website, you should see how many visits you are getting. Review your Analytics and determine if you have enough people visiting your site to warrant the commitment to Live Chat. If you do, then determine how you would manage Live Chat. For every feature you add to your website, you should be able to determine a ROI, keeping in mind that your time is money. How available do you want to be? And in order to answer that question, you should understand how Live Chat works.

# How does Live Chat work?

Once you integrate Live Chat on your website, you can turn it off and on. So while you are working on your computer, you can opt to turn the chat on so that you can engage with visitors on your site.

A visitor on your website is similar to a shopper in a boutique, so asking if you can help them find something can turn a visitor into a customer. Your quick response may spark a conversation and turn a visitor into a client.

NOTES

_____

_____

_____

_____

_____

You can try it out and see if this is something you might want, since many of the Live Chat programs offer a free trial.

Using Olark Live Chat as our example, here is how Live Chat works:

1.  Sign up for an account with Olark.
    http://www.olark.com/

2.  Once you have activated your account, you can install the WordPress plugin on your website and activate it by pasting the code provided by Olark.

3.  Now drag a text widget to your footer and paste the provided code into the widget. For detailed directions, visit the Olark instructions on their website.
    http://www.olark.com/customer/portal/articles/314795-wordpress-integration-guide

4.  You should receive a message that your account is active.

5.  Navigate the options to set your account up. From your dashboard, you can access Settings to set up the different features, which will vary depending on the account you selected. You can change these later, but it's good to know what's possible as you become more comfortable with using Live Chat.

6.  From your Olark Chat dashboard, select whether you are accepting calls. When you are away, the chat box will offer

NOTES

_____

_____

_____

_____

_____

_____

your visitors a message and they can leave you a message.

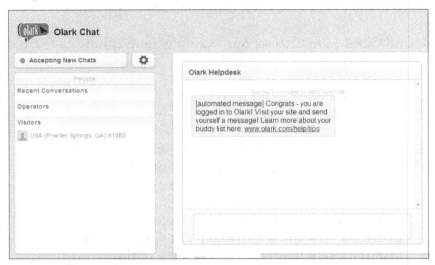

7. You can modify the appearance of your website chat widget which will appear at the bottom of your website.

## NOTES

_____

_____

_____

_____

_____

_____

8.  You can also change the wording that appears in your widget.

9.  As stated earlier, Live Chat offers the opportunity for instant engagement. Testing this functionality over a short period of time will help you determine if this is a good solution for you, or not.

# Live Chat Plugins

There are several different WordPress Live Chat plugins available and like IDX plugins, they each provide different features. Following is a list of some of the features each one offers.

### My Live Chat
Website:     https://www.mylivechat.com
Plugin:      MyLiveChat - Live Chat Plugin for WordPress
Features:    free plan available
             real time traffic monitoring
             customizable widget
             spell check
             email chat transcript for visitors
             typing indicator
             online/offline messaging system
             document sharing
             repeat visitor identification

Notes

## OLARK

Website:        http://www.olark.com/
Plugin:         Olark for WP
Features:       free plan available
                customizable widget
                attention grabber
                greeter
                visitor insights*
                integrations*
                transcripts*
                reports*
                cobrowing*
                activate on smartphone

                *features on upgraded plans

## ZOPIM

Website:        https://www.zopim.com/
Plugin:         Zopim Live Chat
Features:       free plan available
                customizable widget
                iphone application
                integrations
                dashboard analytics
                chat history*

                *length of history varies with plan

NOTES

_____

_____

_____

_____

_____

# Google Tools

Learn how to leverage Google tools to monitor and analyze your WordPress website

- Google Webmaster Tools
- Google Analytics
- Google Places
- Google Plus
- Google Authorship
- Google Hangouts
- Google Keyword Planner
- Google Alerts

Google offers a suite of tools that can help you with your website. This chapter provides an overview of the tools and a basic idea of what each one provides to help you manage your website.

## Google Webmaster Tools

Google Webmaster Tools monitors your site for website performance including indexed pages and crawl status along with site traffic. Much like other Google tools, GWT is about met-

NOTES

_____

_____

_____

_____

_____

_____

rics: what is indexed, what is linked, and what is getting traffic. Since this is a Google tool, it does not include results from other search engines.

The tools offered through Google Webmaster Tools include:

- Submit and check a sitemap
- Generate and check a robots.txt file, and helps to discover pages that are blocked in robots.txt by chance.
- List internal and external pages that link to the site
- Get a list of broken links for the site
- See what keyword searches on Google led to the site being listed in the SERPs, and the click through rates of such listings
- View statistics about how Google indexes the site, and if it found any errors while doing it
- Designate a location

Monitoring your results will help you see which pages have higher CTR (click through rate) and then determine some strategies to improve their rankings. You also may discover that your visitors are interested in topics that you weren't aware of, which will help you in your content strategy.

Once you have set up your Google Webmaster Tools account,

Google                                    Mama Friedman    + Share

Webmaster Tools                              Help    

Home
All Messages (6)        Sort: By site health  Alphabetically        ADD A SITE

## NOTES

_____

_____

_____

_____

_____

_____

---

## Webmaster Tools

You can use the boxes below to verify with the different Webmaster Tools, if your site is already verified, you can just forget about these. Enter the verify meta values for:

Google Webmaster Tools:

Bing Webmaster Tools:

Alexa Verification ID:

---

# Google Analytics

Google Analytics, like GWT, shows total traffic for a website. As SERP becomes more sophisticated, you will see more long-tail search terms which means instead of "real estate in austin tx", you might see "real estate in four points, austin tx" as clicks to your site. Google Analytics may tell you how you were found, which can also help you in your content creation strategy.

As you become comfortable reviewing your analytics, you can expand your strategy based on the results. Create a spread-sheet of keywords that you want to be found with, along with the terms you *are* being found with. Develop blog posts around these terms and monitor the results. You should also leverage this information in creating your local SEO strategy.

**Analyze your competition**. Review the websites and social media presence of your local competition by doing a keyword

NOTES

_____

_____

_____

_____

_____

search and seeing who comes up in the results. You can increase your results by sharing relevant content and adding new content on a consistent basis.

You will need to create a profile for your website on Google Analytics in order to measure your results.

**Account Name**
Accounts are the top-most level of organization and contain one or more tracking IDs.

My New Account Name

Setting up your property

**Website Name**

My New Website

**Website URL**

http:// ▾    Example: http://www.mywebsite.com

**Industry Category** ⑦

Select One ▾

**Reporting Time Zone**

United States ▾    (GMT-08:00) Pacific Time ▾

Once you create the account profile for your website, you will need the tracking information to input on your website.
1. Click on the Admin link on the top right side of your screen. Select Tracking Info from the middle screen.
2. Click Tracking code.
3. Copy the code.
4. Your theme options panel may include a location for this code. If it doesn't, you can use a plugin like Insert Headers

NOTES

_____

_____

_____

_____

_____

and Footers.

5. You will also need to connect your Google Analytics and Google Webmaster Tools. Go back to Property Settings in Google Analytics and scroll down to Google Webmaster Tools.

---

Webmaster Tools Settings

**Webmaster Tools site** optional

If your property is also a verified website in Webmaster Tools, and you are the owner, you can associate your Webmaster Tools data here. Google Analytics will then be able to display some of that data in some reports.

---

# Google Places

*https://www.google.com/business/placesforbusiness/*

As local search becomes more important, it is even more important to confirm your business location with Google through Google Places to be found on Google Search, Maps, Google+, and mobile devices. Since more than 90% of the public searches are for businesses online, this is critical to your visibility and SERP. Once your are verified with Google Places, your clients will be able to post Google Reviews for your business, which is a metric factor in SEO.

Even though you have completed the process of signing up your business with Google Places, you will not be verified until you receive a postcard from Google to verify your business. Once you receive the postcard, you need to input the code on your Google Places profile to verify your business and then your business should begin to appear in local search results.

NOTES

_____

_____

_____

_____

_____

_____

# Google Plus

*http://www.google.com/+/learnmore/*

Create a business page on Google Plus and start sharing information and engaging with other members of the community. Don't make the mistake of excluding this platform from your social media strategy, it is growing in usage and through its features such as Google Hangouts, is increasing in adoption. You also need a Google Plus profile in order to activate Google Authorship, which brings us to the next Google tool you should have.

# Google Authorship

*https://plus.google.com/authorship*

This is a fairly new Google Tool and only requires your activation. Not only does Google Authorship put a face to your content, but when you establish yourself as the original author of the content, Google ranks it higher for you than duplicate content. Your claiming ownership as the author increases its value.

*Google Authorship is all about you.*

Google has started to implement Author Rank which ranks Google Authorship writers based on their engagement and social factors. Early adoption of this program increases your visibility which in turn can impact your Author Rank.

NOTES

_____

_____

_____

_____

_____

# Google Hangouts

*http://www.google.com/+/learnmore/hangouts/*

Through your Google Plus profile, you can create and invite people to Google Hangouts. These are online gatherings where everyone invited can take part in the conversation. You can have live calls with clients, schedule a neighborhood workshop on mortgage programs, collaborate with your team. Google Hangouts works on smartphones, tablets and desktops, and it's even accessible through Gmail.

# Google Keyword Planner

*https://adwords.google.com/ko/KeywordPlanner/Home*

The Google Keyword Tool has been replaced with Google Keyword Planner. The new tool is a little more difficult than previous versions, since with this version Keyword, you'll get historical statistics only for an exact match.

You can search for keywords based on relevant terms and/or landing pages. Get historical statistics and traffic estimates to determine if there is value in using the terms you have selected. You can review the estimated clicks to get an idea of how successful your strategy and/or campaign might be. And if you are considering Google Ads, these estimates can also help you determine a campaign budget.

The best strategy would be to create a local SEO campaign that focuses on relevant long-tail keywords.
• Using zip codes in your content, but still keeping it longtail,

NOTES

_____
_____
_____
_____
_____

so 'waterfront condos in Five Points, Fort Lauderdale Beach 33316' and incorporating this into your content. But don't make it so specific that it is beyond what someone might search for.

- Optimize your business directory listings with the local search terms you are focusing on. Be creative in how you include them.
- Review your social media profiles and optimize them with these keywords as well.
- Review your Google Webmaster Tools and Analytics results to understand what your visitors may be looking for and adjust your content and social media strategy to provide that and increase engagement.
- Consult a professional before you go down the rabbit hole.

# Google Alerts

*http://www.google.com/alerts*

Google Alerts can be helpful in online reputation management as well as content generation. You can create Alerts that let you know what is being said about you. You can also create Alerts to keep apprised of what's being said about topics relevant to your niche.

As you create Alerts, you can determine what you want to know:
- frequency
- type of results
- how many
- where to send results

NOTES

_____

_____

_____

_____

_____

_____

# Using WordPress SEO by Yoast

There are several SEO plugins that you can use on your Word-Press website, but I use WordPress SEO by Yoast, so the following pages will take you through my basic set-up of this plugin. These are the settings *I use* and yours might be different. I strongly recommend that you review the videos and tutorials on the plugin site to familiarize yourself with the options available and make the decision that fits your situation. This information is to illustrate how I use the plugin, and not to dictate how you should use it. *Please review the developer's tutorials.*

1. If you are currently using a different SEO plugin, then you should transfer the data before you deactivate the plugin. And yes, you will need another plugin to do this. SEO Data Transporter can help you with this.
2. Once you complete that step, deactivate the old plugin and delete it so you can install WordPress SEO by Yoast.
3. Go to Settings on WordPress SEO by Yoast. You can click on the tour button if you want to explore the different options of the plugin, or watch the videos offered. You can customize these once you are more familiar with the setup. Again, these screens are just an example of how I use this plugin.

NOTES

_____

_____

_____

_____

_____

_____

The next screen will be Titles & Meta. There are several tabs that you will need to go through to set up your site. So start with General Settings and move through each tab.

### GENERAL
Title settings - the first option is to Force rewrite titles. But this is auto detected and will show whether it needs to be done or note, so I leave it alone.

### SITEWIDE META SETTINGS
One of the important aspects of setting this plugin up is to try and prevent duplicate content from being indexed.So I check the option to **Noindex subpages of archives**. *If you blog a lot, your archive pages will increase and the only page that is important to index is the first one, so checking this will help prevent the additional pages from being indexed.*

The only other thing I check here is **Use meta keyword tags**. There is mixed feedback on this option, but I typically check it. You can choose to select this or not.

Those are the only items that I find important to select on the General Settings tab.

4. Moving on to the Home tab where I set up my home page template. While I do use defaults on this page, I modify what I use. The purpose of using the variables is so that if you ever update the site title or tagline, it will be automatically updated and you won't have to change it manually. Some people convert these variables to numbers, creating a

NOTES

_____

_____

_____

_____

_____

_____

| General | Home | Post Types | Taxonomies | Other |

## Title settings

☐ Force rewrite titles

WordPress SEO has auto-detected whether it needs to force rewrite the titles for your pages, if you think it's wrong and you know what you're doing you can change the setting here.

## Sit⋯e meta settings

☑ Noic⋯ex subpages of archives

If y⋯ want to prevent /page/2/ and further of any archive to show up in the search ⋯s, enable this.

☑ Use meta keywords tag?

I don't know why you'd want to use meta keywords, but if you want to, check this box.

☐ Add noodp meta robots tag sitewide

Prevents search engines from using the DMOZ description for pages from this site in the search results.

☐ Add noydir meta robots tag sitewide

Prevents search engines from using the Yahoo! directory description for pages from this site in the search results.

default amount of characters to use.

Understanding the variables may help you determine what to choose.

%%sitename%% %%page%% %%sep%% %%sitedesc%%
**%%sitename%%** - your sitename
**%%page%%** - since this is your homepage template, it won't have a page, so you can delete this.

NOTES

_____

_____

_____

_____

_____

_____

**%%sep%%** - is a separator, which is usually a dash, but since this is typically determined by your theme, it might be some other type of separator. Since you might change your theme, this could change as well. My choice is to delete the variable and actually use my own separator which is the pipe ( | ) a vertical line. It's your choice.

**%%sitedesc%%** - site tagline or description in your Word-Press General Settings.

So my final choice is:
**%%sitename%% | %%sitedesc%%**

5. I type my site description in the next box and since I selected meta keywords in General Settings, you will see the meta keywords box for me to type keywords in.

| General | Home | Post Types | Taxonomies | Other |
|---|---|---|---|---|

## Homepage

Title template:

%%sitename%% - %%sitedesc%%

Meta description template:

Meta keywords template:

NOTES

_____

_____

_____

_____

_____

If you choose to write your own title and description, just type a number in the boxes.

6.   The next tab has several sections to set up the default Title, Description and Keywords for posts, pages and media. Depending on your sitename, you may just want to use the default %%sitename%% . If your sitename is not very keyword focused or "geolocal", you may want to use the character option here and create a unique title for every page. The caveat here is that you will need to remember to type in a title for every page and/or post you create. The default includes %%page%% which you can remove.

Since I do want my posts to be indexed, I make sure to leave the box for Meta Robots **unchecked**.  Here are my final post options.

NOTES

_____

_____

_____

_____

_____

_____

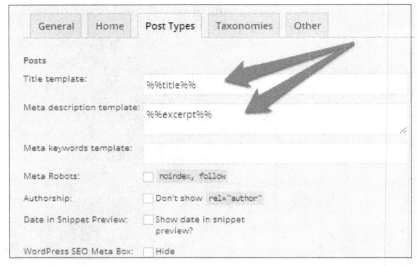

**NOTE**: if you want to use the focus keyword option in the WordPress SEO meta box on the bottom of your post pages, then you can type %%focuskw%% in the Meta keywords template, and this will include this in your search results. Some people agree with this practice, others don't, so again, make your own decision.

7.  Pages will be similar to posts, but some may choose to have their pages set to no index, and then just set options on specific pages through the meta box that appears at the bottom of each page and change the settings on that individual page. Only you can determine the Meta Robots option for your pages, since you know how you set up your pages and the content they contain. I leave it alone.

NOTES

8. Media refers to your images, pdf uploads, etc. which I typically do not want indexed, so the Meta Robots option will be checked in this section.

9. If you are using plugins that offer a custom post type, like events or something else, it will probably be listed on this screen. These are usually an enhancement to your posts or pages and not mandatory to your SERP.

*Remember that anything you check to noindex, nofollow will not be indexed or ranked by Google. This process can be overwhelming, so you might want to get some help with this that is specific to your website and how you have set it up, both with the IDX and the website content.*

## NOTES

10. Since on my website, the Taxonomies would be duplicate content, I select Meta Robots to noindex, follow. The content has already been indexed in its original format, so I don't want it indexed again. An exception to this would be if you design your category pages to include an introduction about the category and then just an excerpt from each page or post.

    To set up the latter option, you would need to create a category template within your website page templates. You should also include a description of each Category and Tag on your website.

You can modify the Title templates as you see fit for your website similar to how I did above. And notice that I included description options for both Categories and Tags. If you do not, then leave the description field blank.

NOTES

_____

_____

_____

_____

_____

_____

Add New Category

Name

*The name is how it appears on your site.*

Slug

*The "slug" is the URL-friendly version of the name. It is usually all lowercase and contains only letters, numbers, and hyphens*

Parent

None ▼

*Categories, unlike tags, can have a hierarchy. You might have a Jazz category, and under that have children categories for Bebop and Big Band. Totally optional.*

Description

*The description is not prominent by default; however, some themes may show it.*

11. The last tab is for the meta data on your pages and posts which I pretty much leave as is.

    NOTE: If your author name is the name of your site, then delete the sitename option from your author. Otherwise, it will show up in search results as Jane Doe Jane Doe.

NOTES

_____

_____

_____

_____

_____

_____

**Categories**

Title template:

%%term_title%% Archives %%page%% %%sep%% %%siter

Meta description template:

Meta Robots:    ☐ noindex, follow

WordPress SEO Meta Box:    ☐ Hide

**Tags**

Title template:

%%term_title%% Archives %%page%% %%sep%% %%siter

Meta description template:

Meta Robots:    ☑ noindex, follow

WordPress SEO Meta Box:    ☑ Hide

**Format**

Title template:

%%term_title%% Archives %%page%% %%sep%% %%siter

Meta description template:

Meta Robots:    ☑ noindex, follow

WordPress SEO Meta Box:    ☑ Hide

**NOTES**

_____

_____

_____

_____

_____

_____

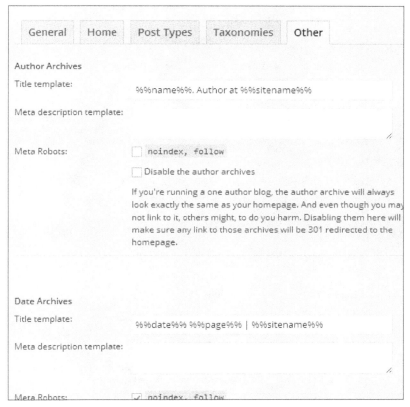

12. The next section is Social, where you can integrate your website with your social media.

    In Facebook, these settings relate to your business page. Although you will have to connect through your personal Facebook account, this is just to approve access to your page. Here are the settings I recommend:

NOTES

_____

_____

_____

_____

_____

_____

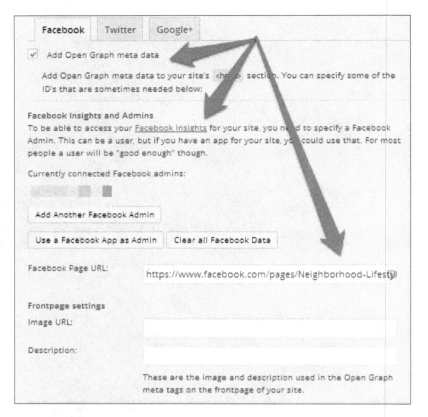

Here are the Twitter settings:

**NOTES**

_____

_____

_____

_____

_____

_____

Google:
Enter your author name, or select from the dropdown. If you have created a Google+ page for your business enter the URL here, and then make sure to go to your Google+ business page and enter your website's URL in the about section.

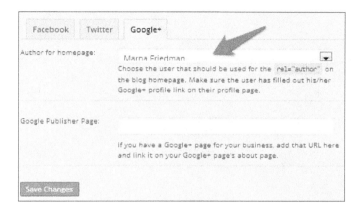

The next section is your sitemap. Here is where you can select the options that appear in your sitemap to be indexed. If you selected no index, follow on categories and tags, then you can select to not include them in your sitemap as well.

NOTES

_____

_____

_____

_____

_____

_____

## XML Sitemap

☑ Check this box to enable XML sitemap functionality.

You can find your XML Sitemap here: | XML Sitemap |

You do not need to generate the XML ~~~~~ nor will it take up time to generate after publishing a post.

## User sitemap

☐ Disable author/user sitemap

## General settings

After content publication, the plugin a ~~~~ atically pings Googl ~ and Bing, do you need it to ping other search engines too? If so ~~~ the box:

☑ Ping Yahoo!
☑ Ping Ask.com

## Exclude post types

Please check the appropriate box below if t ~~ e's a post type that you do NOT want to include in your sitemap:

☐ Posts ( `post` )
☐ Pages ( `page` )
☑ Media ( `attachment` )
☐ Agents ( `myteam` )

## Exclude taxonomies

Please check the appropriate box below if there's a taxonomy that you do NOT want to include in your sitemap:

13. Permalinks

    The first option is a personal preference. Some people like category in the search results, because they think it helps the search engines to know where they are. The next section refers to Stopwords, which are common words some-

NOTES

_____

_____

_____

_____

_____

_____

times not considered by search engines to increase search speed. Since most people use stopwords in their search, I leave this option unchecked. It's your decision. I do recommend selecting Redirect attachment URLs to parent post URL so that attachments will automatically redirect to the post or page where it was attached.

**Permalink Settings**

☐ Strip the category base (usually /category/ ) from the category URL.
☐ Enforce a trailing slash on all category and tag URL's

If you choose a permalink for your posts with .html ...ng else ...a / on the end, this will force WordPress to add a trailing s... ...post pages n...theless.

☐ Remove stop words from slugs.
This helps you to create cleaner URLs by automat... ...oving the stopwords from them.

☑ Redirect attachment URL's to parent post URL.
Attachments to posts are stored in the database as posts, this means they're accessible under their own URL's if you do not redirect them, enabling this will redirect them to the post they were attached to.

☐ Remove the ?replytocom variables.
This prevents threaded replies from working when the user has JavaScript disabled, but on a large site can mean a huge improvement in crawl efficiency for search engines when you have a lot of comments.

☐ Redirect ugly URL's to clean permalinks. (Not recommended in many cases!)
People make mistakes in their links towards you sometimes, or unwanted parameters are added to the end of your URLs, this allows you to redirect them all away. Please note that while this is a feature that is actively maintained, it is known to break several plugins, and should for that reason be the first feature you disable when you encounter issues after installing this plugin.

**Canonical Settings**

Force Transport: Leave default

14. Breadcrumb Settings
    Since most themes offer breadcrumb settings options, I usually leave this page blank and defer to the theme settings.

NOTES

15. RSS

The only change I make on this page is to reword the way I want my content linked in RSS posts. There are explanations of each of the defaults to help you in your selection.

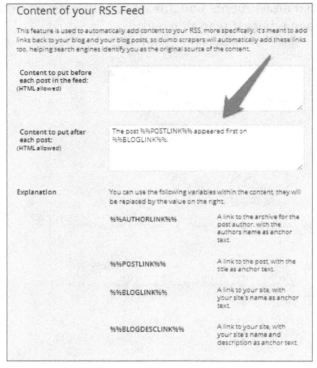

16. Import & Export

This is where you would import from another plugin, and as you can see, the recommendation is to use the plugin we mentioned on page 92 first.

NOTES

_____

_____

_____

_____

_____

_____

If you are importing data, just follow the steps on this page as explained.

17. Edit files
    Unless you know what you are doing, you should not edit these files.

That's it! You have now set up your Google accounts to monitor your site and help you with your SEO. You have also set up the SEO plugin on your website to work with the settings from your Google accounts. Keep in mind that SEO is not a set-it and forget-it program. You will need to continue to monitor your search results and make adjustments or additions as necessary.

The best thing you can do is to blog. Adding fresh, relevant content to your website will reflect activity to the search engines and they will want to know what is going on.

NOTES

_____

_____

_____

_____

_____

_____

NOTES

_____

_____

_____

_____

_____

_____

# Integrating Social Media

Learn how to integrate social media on your real estate website.

- 📖 Growing Your Community
- 📖 Integrating Social Media Platforms

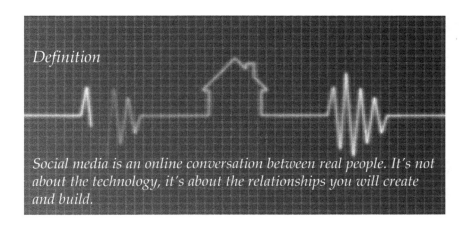

*Definition*

*Social media is an online conversation between real people. It's not about the technology, it's about the relationships you will create and build.*

NOTES

_____
_____
_____
_____
_____
_____

# Growing your community

Even though your website gives you an online presence, you need to augment it with social media. Not every social media application is right for your audience. Understanding how to use social media will help you to engage with your community.

It's important to understand how social media can help with your business strategy. Before you begin to use social media, you need to register for the applications you want to use. Even though you may not use all of the applications listed below, you may want to register for an account to use in the future. If your user name is consistent, it is easier for your community to find you. A quick way to find out if the name you want is available is to use namechk. This application will do a quick search of the name you select and let you know if it is available.

☐ **Namechk**
http://namechk.com/

NOTES

_____

_____

_____

_____

_____

You can type your username in the top box and click OK. Keep in mind that you should use a name that is less than 20 characters, especially since tweets from Twitter are only 140 characters. It's also easier to remember a shorter name. The screen will quickly change to show you which sites have that username available, which names are taken, and also if there are domains with that username available. Clicking on each site will take you to the site, where you can register.

NOTE: *unlocking a technology tidbit for you*

*As you select the sites and register, be sure to remember your user ID and password.*

According to Nielsen, almost 80% of the US internet users frequent social networks and blogs. And close to 40% of social media users access their sites on smartphones - which is another reason why your website needs to be responsive.

# Using Social Media

Most real estate websites include links to their social media on their marketing materials and websites. But they aren't using the social media. Much like a conversation, if you aren't participating, you really aren't involved. And social media is not about your real estate business, it's about your community. You need to be involved with the conversation to engage current and future customers.

NOTES

_____

_____

_____

_____

_____

Social media is about engagement and not sales. While the real estate business tends to focus on listings, participation in social media offers an opportunity to engage your community helping to create and build relationships. Conversations on social media are indicative of what people want to talk about.

Start a conversation and wait for participation. Always remember that this is a conversation, *not a sales pitch*. Sometimes it's great to just listen too. You can learn what people want to talk about.

You can also create networking events so that your online community can meet in person. The earlier you start your networking events, the more opportunity you have to grow your community and increase your business. Invite your clients to participate in the community. They become a representation of your ability to build relationships.

Unlike many other businesses, real estate agents often have social media representation without their being involved. Sites like Yelp may offer reviews and customer interactions without a real estate agent even being aware, especially if they aren't active on social media.

As you can see from the infographic on page 139 from Balihoo, local marketing and participation is an opportunity for all local businesses.

http://balihoo.com/

NOTES

_____

_____

_____

_____

_____

_____

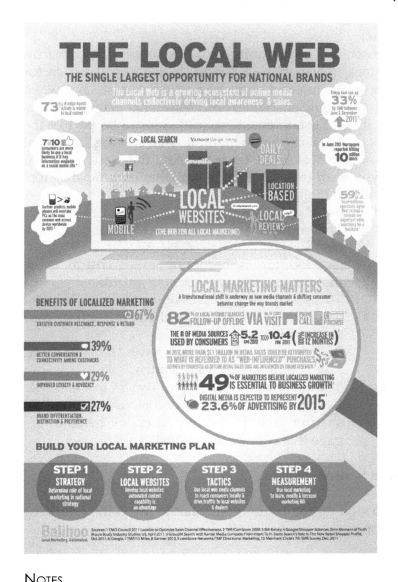

NOTES

# Using Twitter

Twitter presents an opportunity for quick, interactive conversations. A quick explanation of Twitter is: when you search for something on the internet, you can find what exists. But when you search that topic on Twitter, you hear what people are saying about it. Real estate agents should want to be part of the conversation. You also want to monitor what is being said about you, and/or your real estate business. And there are tools that can help you with this. Twitter is available on your desktop, smartphone and tablet.

Name:          This should be your name or name of your business.
Location:      Geo-location is becoming more critical. Note that you can always edit your profile.
Website:       This should be your website so that people know where to go to learn more about you and/or your restaurant.
Bio:           Be informative, you only have 160 characters. This should not be a sales pitch.
Facebook:      You can elect to post your Tweets to Facebook. Keep in mind that not everyone uses Twitter. So sharing your Tweets on Facebook helps to engage that community as well.

*Remember to save changes.*

When you sign up for Twitter, you should input the following information:

NOTES

_____

_____

_____

_____

_____

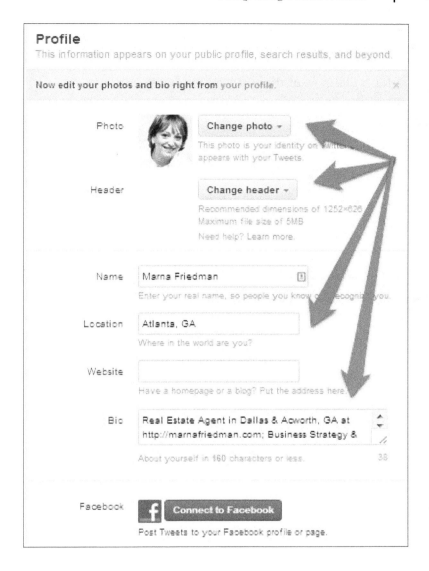

To create a Twitter widget for your website, you can click on the Widget option on the left side of your Twitter profile.

1. Click on the gear at the top of your Twitter screen.

2. Select Edit Profile from the dropdown. And then select Widgets from the left sidebar.

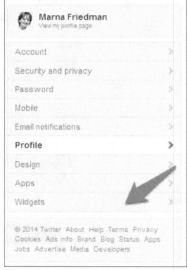

NOTES

_____

_____

_____

_____

_____

_____

3. Make sure to complete all the relevant information as you create your Twitter widget as noted above. You can modify the size and color to coordinate with your website. And the widget will look like the Preview shown above. Once you have completed all of the details, click the Create widget button to capture the HTML code.

NOTES

_____

_____

_____

_____

_____

_____

4. Copy the HTML code in the bottom box to paste into a text widget on your website.

5. Refer to Chapter 10 for instructions on how to use Widgets. Then place a text widget where you want your Twitter widget to appear and paste the HTML code into the widget. Save. View your page, as you can see from the example on page 146.

NOTES

_____

_____

_____

_____

_____

_____

Text

Title:

Chat with us on Twitter

```
<a class="twitter-timeline"
href="https://twitter.com/marnafriedman" data-widget-
id="___  ___  ___  ___">Tweets by
@marnafriedman</a>
```

tElemen

n.twitter

;);}

☐ Automatically add paragraphs

Delete | Close                                    Save

NOTES

_____
_____
_____
_____
_____
_____
_____

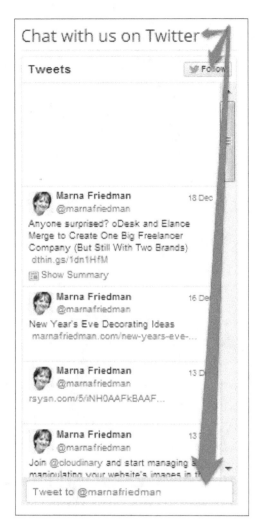

The benefit of using the Twitter widget is that you will not have to worry about updating a plugin or code conflicting. It is also responsive on a responsive website. You will not need all of the OAuth information you probably need for a plugin.

The visitor can elect to Follow you on Twitter, as well as participate in the Twitter stream right from within your website, so engagement is easy.

If you would like to use a plugin instead, there is a list of some that you can use on page 147.

NOTES

Photo:        Upload an image that relates to you and/or your profile. Don't leave this blank, build your brand.

## Twitter Plugins
- ☐ My Twitter Widget
- ☐ Twiget Twitter Widget
- ☐ Twitter

# Using Facebook

While Twitter is about the world, Facebook is about friends and family. As you share information, you will continue to grow your Facebook community. But you need to determine if your community is on Facebook. You can also advertise through Facebooks Ads, and if your community is active on Facebook, a Facebook page might provide a decent ROI.

Some of the questions you need to answer are:
- Are your customers on Facebook?
- Can you engage visitors in a conversation?
- Does Facebook offer a community for your business?
- Will your fan base grow based on content?

You also have the ability to add 13 apps to your Facebook Page Timeline, so you can select apps that are relevant to your business.

Other ways you can leverage Facebook for your real estate business include:
1. Sharing information about the neighborhood.

NOTES

_____

_____

_____

_____

_____

_____

2. Share some listing information
3. Answer questions, address concerns, and acknowledge when people take the time to post on your Page.
4. Update your page with news items and engage friends to ask what they are interested in.

Facebook offers WordPress plugins in addition to those that are available from the WordPress repository. You can read about the options from Facebook at https://developers. facebook.com/blog/post/2012/06/12/facebook-integration-for-wordpress/.

## Facebook Plugins
☐ Facebook Social Plugin Widgets
☐ Facebook Widget
☐ ThemeLoom Widgets

# Using Pinterest

Pinterest is one of the fastest growing social media platforms. According to comScore, Pinterest was the fastest independent site to hit 10 million monthly unique visits in the U.S. As participation in Pinterest continues to climb, businesses are also quickly discovering ways to connect with individuals. Pinterest is all about visual sharing.

Let your community know that you are on Pinterest and to share photos by integrating it onto your website. Share photos from listings, events around town, new products, decorating ideas, anything to engage current and future customers.

NOTES

_____

_____

_____

_____

_____

_____

So how do you use Pinterest?
- Think visual
- Organize your boards - consider your audience
- Be descriptive so your visitors are compelled to repin
- Use relevant keywords to be found easily
- Engage, don't sell

To join Pinterest, you just need to register at the Pinterest website. In order to create a widget, you will need to join Pinterest as a Business.

1. Go to https://www.pinterest.com/business/create/
2. Complete the information requested, and click Create Account.

The first time you login to Pinterest, you should familiarize yourself with what is possible. Click on the About button in the upper right corner, and navigate through all of the sections.

### General Information About Pinterest
- Organize your boards by interests, with one interest per board - this makes it easier for viewers to follow
- Update your boards regularly - so only create as many as you can maintain
- Use a call-to-action on your pins to engage others
- Add text to images, make your pins descriptive and targeted
- Use searchable keywords in board names
- Select a compelling image for your board cover

NOTES

_____

_____

_____

_____

_____

_____

Like Twitter, Pinterest offers the ability to create widgets for you to add to text widgets on your WordPress website with a little modification of the code provided, as seen below.

1. Click your profile name in top right corner of Pinterest. Scroll down to Visit Help Center and click to learn what's possible.

2. Type WordPress Widget in the search box.

NOTES

_____

_____

_____

_____

_____

_____

3. You will have three options to select from and should explore all three.

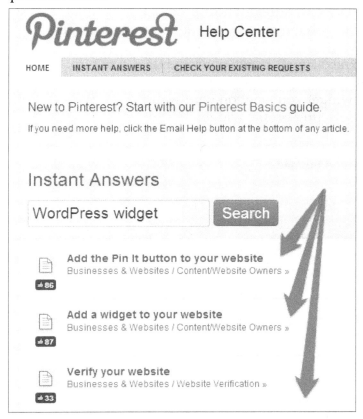

The third option, verify your website, is probably where you should start. Once you verify your website, you will also have access to analytics, which can help you in learning how your Pinterest activity is doing.

NOTES

_____

_____

_____

_____

_____

_____

Pinterest offers several different ways to verify your website, as well as directions to walk you through the process with several different host providers. Complete this step before you move forward with creating a widget.

4.  The next step is create your widget. Go to http://business.pinterest.com/widget-builder/#do_pin_it_button and select the widget you want on your website. For this example, I am selecting profile widget.

5.  Click on your Pinterest profile on the right hand side of your screen to copy your URL.

NOTES

_____

_____

_____

_____

_____

6.  Paste the URL in the Pinterest URL box and select your widget option. You can modify size later.

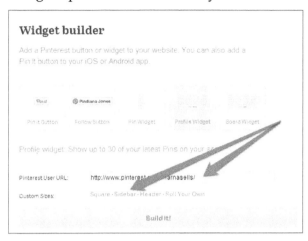

You can modify the dimensions if you like. And then click Build it!

7.  Below your preview, you will see code. You can either go back and choose Roll Your Own, or modify the size within the code - but only change the measurements!

8.  You will actually need to modify the top code a little here since the javascript provided will not work in a text widget, and you will need the link code as well. But if you click the help link at the bottom of the page, it will direct you to a page where you can copy code. See the steps and final code on page 154.

NOTES

_____

_____

_____

_____

_____

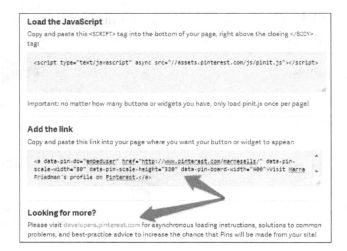

**Load the JavaScript**

Copy and paste this <SCRIPT> tag into the bottom of your page, right above the closing </BODY> tag:

```
<script type="text/javascript" async src="//assets.pinterest.com/js/pinit.js"></script>
```

Important: no matter how many buttons or widgets you have, only load pinit.js once per page!

**Add the link**

Copy and paste this link into your page where you want your button or widget to appear:

```
<a data-pin-do="embedUser" href="http://www.pinterest.com/marnasells/" data-pin-scale-width="80" data-pin-scale-height="320" data-pin-board-width="400">Visit Marna Friedman's profile on Pinterest.</a>
```

**Looking for more?**

Please visit developers.pinterest.com for asynchronous loading instructions, solutions to common problems, and best-practice advice to increase the chance that Pins will be made from your site!

9.  Go to the Developer's link to copy the JavaScript code you will need.

If you'd like to load the JavaScript asynchronously, you can do so by copying and pasting this snippet anywhere on your page:

```
<script type="text/javascript">
(function(d) {
    var f = d.getElementsByTagName('SCRIPT')[0], p = d.createElement('SCRIPT');
    p.type = 'text/javascript';
    p.async = true;
    p.src = '//assets.pinterest.com/js/pinit.js';
    f.parentNode.insertBefore(p, f);
}(document));
</script>
```

You only need to load this script once per page, no matter how many buttons or widgets you use.

## NOTES

_____

_____

_____

_____

_____

_____

10. Copy the code, being sure to use it only once per page! Add the Link code below that code and copy all of that code into your text widget where you want it to appear as below.

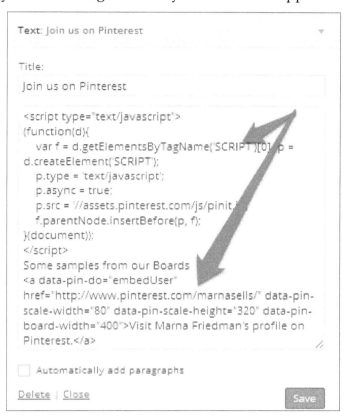

Text: Join us on Pinterest

Title:

Join us on Pinterest

```
<script type="text/javascript">
(function(d){
    var f = d.getElementsByTagName('SCRIPT')[0], p =
d.createElement('SCRIPT');
    p.type = 'text/javascript';
    p.async = true;
    p.src = '//assets.pinterest.com/js/pinit.
    f.parentNode.insertBefore(p, f);
}(document));
</script>
Some samples from our Boards
<a data-pin-do="embedUser"
href="http://www.pinterest.com/marnasells/" data-pin-
scale-width="80" data-pin-scale-height="320" data-pin-
board-width="400">Visit Marna Friedman's profile on
Pinterest.</a>
```

☐ Automatically add paragraphs

Delete | Close                                    Save

11. If you want to feature individual boards, you can create a widget for each one. Add a description line before the link code. Paste link code for each board, but only use JavaS-

NOTES

_____

_____

_____

_____

_____

_____

cript code once. Save the widget and view.

Join us on Pinterest

Some samples from our Boards

Marna Friedman

See On *Pinterest*

The great thing about using the Pinterest widget is that if you have created boards for different niche markets, you can include the board on those landing pages of your website. Then of course, there are also the widgets that Pinterest offers for you to try, and WordPress plugins to integrate Pinterest on your WordPress website.

NOTES

_____

_____

_____

_____

_____

_____

## Pinterest Plugins
- ☐ Pinterest Master
- ☐ Pinterest Pinboard Widget
- ☐ WordPress Canvas Widgets

*Of all of our inventions for mass communication, pictures still speak the most universally understood language.*

*Walt Disney*

# Using YouTube and Flickr

Videos and photos also add another visual component to your business. You can create a YouTube page for your website and share the videos from your business there. This can also link back to your website, Facebook, Pinterest, LinkedIn and you can Tweet about it.

Ways to use videos and photos:
- Add videos to website
- Share tidbits of information for your community
- Create videos of DIY home projects, neighborhood events, and more.
- Interview customers about their real estate experience
- Interview local business owners to share information

NOTES

_____

_____

_____

_____

_____

_____

about the area

To incorporate YouTube and Flickr on pages and widgets, you can use the oEmbed function explained in chapter 5 on page 64.

For a widget of your Flickr photos though, you will need a plugin.

### Flickr Plugins
- ☐ Fast Flickr Widget
- ☐ Flickr-stream
- ☐ Slickr Flickr

# Using LinkedIn

LinkedIn is a valuable tool for connecting with like-minded professionals. You can quickly create a group on LinkedIn and begin to have discussions on relevant topics. Building a community on LinkedIn offers an opportunity to share information and ask questions of decision-makers who in turn can support your business.

As a professional group on LinkedIn, you are also interacting as an individual instead of a brand, making it more personal. Prepare some relevant discussions before launching your group. Also participate in other groups that are relevant to your brand and business. Always keep in mind that you should be focused on sharing information and building a community, not about selling. While you are participating in other groups, learn who

NOTES

_____

_____

_____

_____

_____

_____

the key influencers are and invite them to connect with you.

Review your current network and determine if some of your connections can introduce you to others. Understanding how to leverage the community you already have will help in continuing to help it grow.

Join local business groups and network with other people in your area. This can be a great way to share information and build lead generation. LinkedIn has significant opportunities to gain exposure for real estate professionals as a location for networking and possible relocation customer opportunities.

Like Twitter and Pinterest, LinkedIn also offers a widget that you can integrate onto your website. You can choose to show or hide connections. To create a LinkedIn widget for your website, go to https://developer.linkedin.com/plugins/member-profile-plugin-generator.

## NOTES

_____

_____

_____

_____

_____

_____

And you can see the preview as you select options.  Click Get Code and copy the code to paste into a text widget to place in your selected location on your website.

NOTES

_____
_____
_____
_____
_____
_____

# Social Share

Learn how you can extend your message beyond your website and engage your community
- 📖 Sharing content
- 📖 Engagement
- 📖 Getting them back to your website

If you haven't heard it yet, here it is - if you want to become known as an expert and grow your business, you need to share information and engage your visitors. The easiest way to do this is through blogging.

## Sharing content

Many businesses, even real estate businesses, think that having a Facebook page and posting new listings is social sharing. Let me tell you now that this is wrong, in so many ways. First of all, blasting about your listings is not sharing. And the only people that may like your post are friends and the seller. But if you create a blog post about an amazing home and how the seller did something unique, you can not only share the listing on Face-

NOTES

_____

_____

_____

_____

_____

_____

book, but you now shared something that visitors might find interesting. The listing is incidental, the story is about what the seller did that was different and how other homeowners might try it or, even better, be interested in buying the home.

Before you create great content, you need to have all of your tools in place. So having share buttons on your posts is important. The easier you make it for people to do something, the more likely they are to do it.

## How to incorporate social share

Your theme may include social share buttons. If it does, I recommend you use them since adding a plugin when one isn't necessary can slow your site down. But if you still want one, then here are some options.

- [ ] Cunjo: The Best Free Social Share Plugin
- [ ] Hupso Share Buttons for Twitter, Facebook & Google+
- [ ] Social Login, Social Sharing, Social Commenting and more!

You will need to go through the set-up process and determine which platforms you want included in your share options. There are so many social media platforms and adding all of the options will overwhelm your visitors. But not including some options may limit sharing of your content. This is where you need to consider your audience, and your own search results.

Some of the share plugins offer analytics which might be help-

Notes

_____

_____

_____

_____

_____

_____

ful in seeing which posts create engagement. But here again, you need to consider what you can manage. All of the analytics can be overwhelming and divert your attention. If you are creating relevant content and sharing it, the engagement will happen.

Some of the plugins will offer share buttons with counts in them reflecting how many times a post was shared on a particular platform. This is great for large sites, but probably not for a niche site. So you should select icons that reflect the platform, but without numbers.

Another item you might be able to customize is where your social icons appear. There are several places where this can be including at the top of your post under your title or at the bottom when the article ends just before comments. And then there are widgets that might appear on the right or left hand side of your website. I prefer my social share buttons at the bottom of my blog posts. It's a preference. Is sharing easy and intuitive for them?

I like the large callout to Share before the buttons on Hupso, as well as the option to email or print a post. And I limit the share platforms to the five above because I believe those are the ones my visitors use. If I was in a different industry, perhaps

NOTES

_____

_____

_____

_____

_____

_____

the list of sites would be different because my visitors use different platforms.

## How you share

Now that you have the share buttons for your visitors to share, you need to have a plan so that you can share your content. As you develop your website content, you will want people to read it. So the more exposure you can get for your content, the better the opportunity for it to be seen. Sometimes this can be just another task to get done, unless you automate it.

Adding an autoposting plugin to your WordPress website will automate the process of sharing your content. Some of them will be more involved to set-up, so be prepared. Keep in mind that the additional installation steps are for your protection. I recommend that you read through all of the documentation and review the screenshots to help you make a decision.

Jetpack by WordPress.com offers a module on autopublish among its many different features. This might be a good place to start, since you can also enable some other functionality with the same plugin including site analytics on your dashboard.

Here are some autopublish plugins you might use:

- Jetpack by WordPress.com
- NextScripts: Social Networks Auto-Poster
- Social Media Auto Publish

NOTES

_____

_____

_____

_____

_____

Once the plugin is set-up, it will autopublish your posts to the social media platforms you have selected. This all sounds great right?

**Tips on sharing strategies[1]:**
1. Posts with 80 characters or less get 60% more engagement.
2. Question posts get 100% more comments.
3. Photos get 53% more Likes, 104% more comments, and 84% more click-throughs

# Timing

Understanding when people use social media can be an important part of your content and sharing strategy. Think about where your visitors are, or who you want your visitors to be. If they are like most people, then they use social media during their breaks from work, before or after work. If you post at 10a, then you are probably not reaching them. And since WordPress offers the ability to schedule your posts, which in turn would schedule your sharing, then you need to be strategic in when this occurs.

**Tips on timing[1]:**
1. Focus on **when** you post: Facebook peak activity is at 3pm
2. Facebook activity peaks on Wednesday
3. Posting 1-2 a day gets 40% more engagement
4. Posting 1-4 times per week gets 71% more engagement

[1]http://blog.kissmetrics.com/more-likes-on-facebook/

NOTES

_____

_____

_____

_____

_____

_____

While the statistics above give you a basic idea of when and what to post, these are statistics, and may not be relevant to your community. Be strategic in the timing of your posts to see if it impacts your engagement.

# Engagement

The most important part of sharing is to engage with your visitors. The timing of your posts and your sharing may be consistent with the statistics, or unique to you. The only way you will learn this is by testing. Test the timing of your posts, test the way you create your message, and also look at how you engage. What gets you to click on a link? What gets you to share a post?

Another strategy would be to look at your competition. Are they blogging? What is their engagement? What content is relevant to readers? Visit the syndication sites and see what content they are sharing. Is it relevant to your niche?

Things that can effect engagement
- Photos create interest and offers an opportunity for your content to be virally shareable

- Regular blogging increases engagement

- Provide meaningful content

NOTES

_____

_____

_____

_____

_____

_____

# Getting them back to your website

All of the sharing is great but the objective is to get them back to your website. How many times have you read a great article and then visited the author's website to see what else they have shared? That's who you want to be. A visitor is probably not going to scroll through your Facebook posts. But they might look at other articles on your website. The organization of your content needs to be easy for them to navigate.

While your content is being autopublished to social media sites, you should engage in Facebook and Google groups participating in discussions. Find groups that are relevant to your niche and contribute to the conversation. Share information beyond what you have posted in social media sites. Offer reasons for them to visit your website. But remember to put the megaphone away. Build a reputation as an expert in your niche, become a resource. And don't forget to listen. Listen to the other conversations that are going on. What do people want to know and how you can help them. These groups can not only help you with engagement, but can also help you with content ideas.

As you share information on your website, redirect visitors there. If you wrote an article about kitchen organization that featured a special product, don't send visitors to the product website, send them to your article. Sometimes you become so intent on sharing, that you forget that little detail.

NOTES

_____

_____

_____

_____

_____

## NOTES

_____

_____

_____

_____

_____

_____

# Email

Learn how to expand the functionality of your WordPress website

- 📖 Email Plugin
- 📖 Creating a List
- 📖 Email Newsletters

Not everyone uses social media. So how do you engage with them? A good option is email.

## Email Plugin

An email newsletter offers you the opportunity to share relevant information with your clients while building trust and your reputation. How can you do this?

Create an email newsletter template that includes:

- Local news or current events
- New real estate in your area
- Local school district statistics
- Focus on a local business - restaurant, store, etc.

NOTES

_____

_____

_____

_____

_____

_____

Real Estate Detail

*Farming to email addresses that did not opt-in is considered SPAM – bottom line.*

The first step in your email strategy is to offer an email list sign up on your website. Before you install any plugins, you need to select an email provider for your list. Some of the real estate CRM programs offer an email list component.

## Current Database/CRM

Find out if your CRM offers an email marketing program. If it does, then find out if they offer a WordPress plugin to incorporate it into your website, and if not, do they offer HTML widget code that you can place in a widget on your website?

**NOTE**: Keep in mind that this database should include people that have opted in to email. If not, then emailing them may be considered SPAM, and you shouldn't do it.

If you do not have a current list, then decide where you want to create your list. Do you want to use a CRM or do you want to use an email program? If you select an email program, can you capture leads into your CRM and email list at the same time?

I use Gravity Forms with MailChimp. So in addition to the

NOTES

_____

_____

_____

_____

_____

_____

email signup on my website, my contact forms also feed my leads to MailChimp.

Some of the CRM programs that you can feed leads to from your WordPress website include:

- ☐ **CONTACTUALLY**
  https://www.contactually.com/
- ☐ **REALTY JUGGLER**
  http://www.realtyjuggler.com/
- ☐ **ZOHO**
  http://www.zoho.com/

If you are not using a CRM, and even if you are, you can still choose to use an email marketing platform like MailChimp for your email marketing campaign. Some of these companies include:

- ☐ **AWEBER**
  http://www.aweber.com/
- ☐ **CONSTANT CONTACT**
  http://www.constantcontact.com/
- ☐ **MAILCHIMP**
  http://mailchimp.com/
- ☐ **MADMIMI**
  https://madmimi.com/

# Email Plugin

Once you select the email platform you want to use, you can usually create HTML code within your account that can be used to create a sign-up widget on your website. You can also

NOTES

_____

_____

_____

_____

_____

search the plugin repository to see if there is a plugin for your email provider.

Once you create your sign-up widget, test it with your own email to ensure that it works. This will also put you on your email list so that you receive a copy of every newsletter you send out.

# Creating a List

Think about your email strategy before you create your lists. The signup on your website may be able to offer options to your visitors. Think about it. A buyer is probably looking for different information than a seller. So offering an option, or the ability to subscibe to both might be a good idea. If you are creating your first list, this is a great time to map out your strategy. If you already have a list, you will need to determine how you will use it. Keep in mind that you can always have a BUYER call out and SELLER call out in your newsletter.

What can you manage? This is a major decision factor in setting up your email newsletter program. If you offer a quarterly newsletter, you will need to create one. You are building a website and online presence, and not everything needs need to be built at the same time. An email signup can be added later.

NOTES

_____

_____

_____

_____

_____

_____

# Email Newsletters

Depending on the email platform you selected, they will probably provide email templates that you can customize with your logo and colors. Determine what you want to include in your newsletter, and design your template around that. The purpose of your email newsletter is to:
- Establish yourself as an expert
- Increase your referrals
- Create calls to action

Make a list of the content you would like in your newsletter and then use that list to create a layout. The layout will help you decide on a layout.

Your content list might include:
- Home selling tips
- Home buying tips
- Home maintenance
- Market trends
- Local focus
- Success stories
- Featured listing

Remember that your email newsletter is part of your branding. Look for a template that is consistent with your website and other marketing materials.

NOTES

_____

_____

_____

_____

_____

_____

## NOTES

_____
_____
_____
_____
_____
_____
_____

# Content Strategy

Learn how to increae the engagement of your WordPress website

## Content Strategy

If you haven't gotten the message yet, let me tell you that blogging is very important to your website strategy. And one of the biggest benefits of using WordPress for your website is the blogging platform. It's ease of use along with the features it offers, will help you with your content strategy.

The first step in your content strategy is coming up with a list of topics you are already comfortable blogging about that would add value to your visitors. Another factor is the ability to schedule posts. So writing a blog post every day is not necessary. You can allocate a day and time for blogging and then schedule when you want them to post.

You should review your keywords and your niche focus to help

NOTES

_____

_____

_____

_____

_____

_____

you with your list of topics.

Remember that a job worth doing is worth doing well. Creating content that interests your readers and provides value will help to increase your value to them and build your reputation. They will want to come back to learn more.

## 5 Steps to Creating a Good Content Strategy

1. **Create titles that evoke interest.** You talk to buyers and sellers every day. Think about the questions they ask you, because there are probably great topics for blog posts in those conversations.

2. **Include SEO keywords**. Try to determine how to include your keywords in blog post titles and content. This can sometimes become the focus more than the content, and when it does you need to step away. The article should be about valuable content for your readers, not about SEO.

3. **Include photos**. You've heard it before, a picture is worth a thousand words. Remember that this photo will appear when you autopublish your post on Facebook, Google+ and any other platform you share it with. So it needs to be relevant to the headline and content of your post.

4. **Short and sweet**. No, not your blog post, your sentences. Keep the sentences short to make it easier for your readers. Use lists and/or bullets, photos and white space to make it easier for your readers. A blog

NOTES

_____

_____

_____

_____

_____

_____

# HEADLINE IDEAS

### SHARE INSIDER KNOWLEDGE
*Have you heard about ...? | The secret of ...*

### PROBLEM SOLVING
*Here's a Quick Way to... | Get Rid of (problem) ...*

### CURIOSITY
*What Everyone Should Know About (topic) ...*

### TIPS
*Top 10 Tips to ...*

### MISTAKES
*5 Mistakes to Avoid When (issue)...*

NOTE: *unlocking a technology tidbit for you*

*Remember not to confuse pages and posts. Pages are for static content and posts are newsworthy items. Understanding the difference can not only be key to your content strategy, but can impact your SEO.*

NOTES

_____

_____

_____

_____

_____

post should convey a message while being easy and quick to read. You don't want to lose their interest.

5. **Call to action**. This should not be a sales pitch. In fact, it doesn't have to be a visual call to action. It can be as simple as ending your post with a question. Ask your readers for their input, their ideas, ways they solved a problem. Engage them by asking a question in the hopes of getting a response. If you do get a response, make sure you join the conversation even if it's just to say thank you. This acknowledges the reader's effort in responding.

# Types of Blog Posts

There are several different types of blog posts, but if you break them down into categories, you can use them to help develop your editorial calendar. And then as you grow your following, you can determine which type of blog post creates the most engagement.

## How-To Post

This is a great post type for real estate, and usually is an easy read. It is similar to the chapter on SEO where you were given the steps on how to set up a plugin along with screen captures.

- How to Organize a Pantry
- How to Create a Floorplan
- How to Get Your Home Loan Approved

NOTES

_____

_____

_____

_____

_____

## List Post

List posts are just that - lists. And they are successful in creating interest. They offer a promise of something - 7 ways to, 5 reasons, 10 tips. And that is impactful. The number doesn't matter, it's just that it's a list. If the list is long, you can break it down to more than one blog post. 10 ways to .... 5 more ways to.

- 5 Ways to Finance Your Home Purchase
- 10 Reasons to Live in Chicago
- 6 Secrets of a Successful Home Sale

## Curated Post

Part of becoming known as an expert is your ability to share information learned from other people. A curated post would be an example of this. Your office may have just had someone come talk to you about HUD programs. Create a curated blog post on what you learned, sharing links to the HUD website and highlight some of the programs that are available. You don't need to write a lot of content, because you can curate the content that already exists. A curated blog post can include video, infographics, etc. that help you share relevant information.

- New Home Ownership Funding Programs
- Business Growth as " " Announces New Location
- Shopping Center Attracts New Tenants

# Editorial Calendar

Once you understand how to create a content strategy and the different types of blog posts, you should be able to come up with a list of topics that you are ready to share with your readers. I have found that an editorial calendar is a great way to

NOTES

_____
_____
_____
_____
_____
_____

organize my topics and help me execute my strategy. The great thing is that there are editorial calendar plugins that you can add to your website to help you keep everything in one place. If you have posts scheduled, and something comes up that you really want to blog about right away, you can just drag your scheduled post to another day and add your new post.

## Editorial Calendar Plugins
- ☐ CoSchedule
- ☐ Edit Flow
- ☐ Editorial Calendar

## Scheduling a post:
Here is how you schedule a post in WordPress:

1. Click Edit.
2. Change the date.
3. Change the time.
4. Click OK.
5. Save by clicking on Schedule.

NOTES

_____

_____

_____

_____

_____

# Using Photography

Learn how to use photography in your WordPress website
- 📖 What You Need To Know
- 📖 Finding Photography
- 📖 Editing Photographs

As a real estate agent, one of the most important aspects of your website will be the photographs. Understanding some basic rules and how to use them will be very beneficial.

# What You Need To Know

Creating original content for your website and blog can be a challenge, but it's nothing compared to the photographs. It can be a challenge to find the right photographs to use on your home page or neighborhood pages. And then finding just the right image to use for a blog post can take longer than writing it. But that photo is not yours. You can't use it, unless you have permission. If you do have permission, get it in writing.

But the easier thing would be to take your own photos. If you

NOTES

_____

_____

_____

_____

_____

_____

take the photos, then there is no question about your right to use them.

## Website photography tips:

- Do not use photos from other websites without permission, *in writing*

- Use your own photos

- Do not use photographs of listings that are not yours

- Do not take photographs of homes that are not your listings

- If you are using professional photographs, make sure you keep copies of your rights to the photos and watermark them

- If you use a photo service, keep copies of your agreements

- Ensure that you are adhering to the Fair Housing Law

You can also watermark your photos to claim them as yours and monitor whether anyone is using them. There are Word-Press plugins that you can use to watermark your images.

- ☐ Easy Watermark
- ☐ Image Watermark
- ☐ Watermark WP Image Protect

NOTES

_____

_____

_____

_____

_____

_____

Services that can help you with protecting your photographs:

- ☐ Digimarc
  https://dfi.digimarc.com/

- ☐ PicMarkr
  http://picmarkr.com/

- ☐ Image Raider
  http://www.imageraider.com/

And just in case you thought Pinterest was a good source of images, think again. You cannot download images and upload them to your website to use as your own. And pinning images to Pinterest is a completely different topic than using Pinterest images on your website.

## Fair Housing and Photography Use

The Fair Housing law says...:

*As made applicable by section 803 of this title and except as exempted by sections 803(b) and 807 of this title, it shall be unlawful--*

*(c) To make, print, or publish, or cause to be made, printed, or published any notice, statement, or advertisement, with respect to the sale or rental of a dwelling that indicates any preference, limitation, or discrimination based on race, color, religion, sex, handicap, familial status, or national origin, or an intention to make any such preference, limitation, or discrimination.*

*(d) To represent to any person because of race, color, religion, sex,*

NOTES

_____

_____

_____

_____

_____

_____

*handicap, familial status, or national origin that any dwelling is not available for inspection, sale, or rental when such dwelling is in fact so available. and is not for the exclusive use of one such group.*

Besides adhering to EHO laws, your website should also include the Fair Housing logo on every page which you can achieve by placing it in a footer widget.

Here is where you can find the EHO logo: http://portal.hud.gov/hudportal/HUD?src=/library/book-shelf11/hudgraphics/fheologo

# Finding Photography

There are several websites that offer subscriptions for royalty free photographs to use on your website. You can also hire a professional to take photos for you, or you can take them yourself. *If you hire a professional, confirm that you will have copyright of the images.* And then there are websites that offer free photography for you to use, but you need to check the Creative Commons and copyright laws.

Here is a list of websites that offer free images for you to use on your website:

☐ **Compfight**
http://compfight.com/

☐ **Death To The Stock Photo**
http://join.deathtothestockphoto.com/

NOTES

_____

_____

_____

_____

_____

_____

- [ ] **Everystockphoto**
  http://www.everystockphoto.com

- [ ] **Flickr**
  http://www.flickr.com/creativecommons/

- [ ] **Fotor**
  http://foter.com/
  WordPress Plugin:     Free Stock Photos Foter

- [ ] **Google Image Search**
  http://www.google.com/advanced_image_search

- [ ] **Pixaby**
  http://pixabay.com/

- [ ] **Morgue File**
  http://www.morguefile.com/

- [ ] **Stockvault**
  http://www.stockvault.net/

- [ ] **Unsplash**
  http://unsplash.com/

- [ ] **Wikimedia Commons**
  http://commons.wikimedia.org/wiki/Main_Page

NOTES

_____

_____

_____

_____

_____

# Editing Photos

If you are using photos from your smartphone on your website, you should edit them before you upload them to your site. These files are usually very large and higher quality than you need on your website. In addition to editing them before you upload them to your site, you can edit them even more from within the WordPress media uploader as well.

Once you upload an image, you can click edit to modify it.

You can crop the image by clicking on the crop tool and selecting a starting point within the image

A dotted frame will appear over the image for you to select the area you want to keep.

Save.

NOTES

_____

_____

_____

_____

_____

_____

You will see that the screen will brighten the area you want to keep and shade out the part you don't want. The dimensions of the image you want will also appear on the right side of the screen.

You can also scale the image as you can see on page 188. You only need to input one number, because the other number will be proportional to the first, and it doesn't matter which number you select. Save your changes.

NOTES

_____

_____

_____

_____

_____

_____

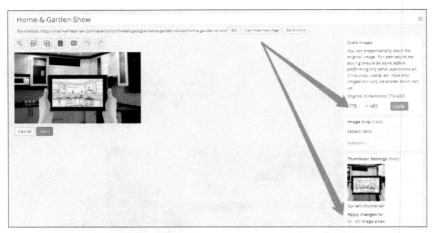

In addition to these edits, you can also edit the image once you place it on your site with borders and other CSS classes.

You can change the alignment of the image, modify the link, add a Title, Caption or any of the boxes listed.

Save your changs.

Click on Advanced Settings

## NOTES

_____

_____

_____

_____

_____

_____

Advanced Settings offers you even more image editing options.

You can add a border to your images, change the border color and modify the offset space. The border can stylized also with standard CSS classes.

NOTES

_____

_____

_____

_____

_____

## NOTES

_____

_____

_____

_____

_____

_____

# Website Optimization

Learn how to improve the performance of your WordPress website
- Testing your website performance
- Using a CDN
- Using W3 Total Cache

NOTE: *unlocking a technology tidbit for you*

*"Fast is better than slow"*
- one of Google's 10 things they know to be true

## Testing Your Website Performance

One of the things that can hurt your search engine results and cause visitors to leave your website is the speed at which it loads. As you add content, images, pdf files, and plugins it begins to impact your website performance. Fortunately, there are ways to test this and react if needed.

NOTES

_____

_____

_____

_____

_____

_____

# Browser Testing

Before you check for performance, you should test your website in each browser to make sure that the results are consistent. This includes Chrome, IE8 or greater, Firefox and Safari. If your site is responsive, you should also view it on an iPad and smartphone.

You can test for responsive features here:

☐ **Browserstack**
http://www.browserstack.com/responsive

☐ **Responsinator**
http://www.responsinator.com/

# Website Performance

There are certain things you can do to quickly to enhance your website performance. You can implement these before or after you test your website performance.

1. **DELETE YOUR SPAM COMMENTS.**
Even though you have a SPAM prevention plugin, there are probably still SPAM comments on your site. Checking this regularly and deleting them is a good maintenance practice,

2. **DELETE UNUSED PLUGINS**
As you are setting up your website, you might be installing plugins, testing them and then deciding you don't like how they work. But you really need to deactivate and delete them. Since they are still activated on your site, they could be impacting your website performance.

NOTES

_____

_____

_____

_____

_____

_____

3. **DELETE POST REVISIONS**
Every time you modify and save a page or post, Word-Press creates a revision. If you are editing a post, it would be a good practice to preview your changes as you are working and save when you are done. If you are more comfortable with saving as you modify, then be prepared to clean up post revisions on a fairly regular basis. To do this you will need to use a plugin,

☐ Revision Cleaner
☐ Thin Out Revisions
☐ WP Document Revisions

You can also use ManageWP, which is a premium plugin that offers this functionality with the click of a button.

4. **CLEAN-UP DATABASE**
As you continue to modify your website, adding content and more, you are adding more "junk" to your database. So a periodic clean-up will help to improve your website performance. And since you are not the only one with this problem, there are plugins that can help you with this.

☐ WP Clean Up
☐ WP-DB-Backup
☐ WP-Optimize

As we have stated before, any time you install or delete plugins and/or themes, be sure to back-up your site before-hand in case you need to restore your site.

NOTES

_____
_____
_____
_____
_____
_____

5.  **CACHE YOUR SITE**
    Caching your website can significantly improve your website performance and I will provide directions on how to use two of these later in the chapter. *Note that using one of these plugins will probably replace you needing to update your htaccess file.*
    ☐ Quick Cache
    ☐ W3 Total Cache
    ☐ WP Super Cache

## Testing Your Site

So now you are ready to test your site. A website performance test should provide information on site speed as well as page sizes and response. The test should also include recommendations on ways to improve your website performance.

The website performance site I use is GTMetrix because they provide a WordPress Optimization guide and a WordPress plugin to help me monitor my website. But there are others:
☐ **GTmetrix**
   http://gtmetrix.com/
☐ **Pingdom**
   http://tools.pingdom.com/fpt/
☐ **WebPageTest**
   http://www.webpagetest.org/

   The YSlow browser extension can also be used for page speed, which is available at http://developer.yahoo.com/yslow/. The Google Page Speed is available at the

NOTES

_____

_____

_____

_____

_____

_____

Chrome Store (https://chrome.google.com/webstore/search).

Once you run your performance test, you will know where there are opportunities to improve your site's performance. While it may seem strange, you will need to install more plugins to improve your website speed and performance.

# Using a CDN

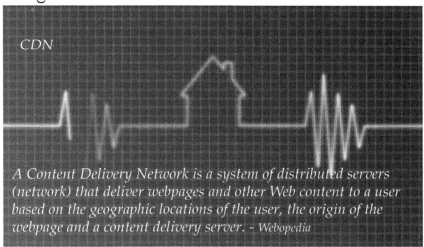

CDN

*A Content Delivery Network is a system of distributed servers (network) that deliver webpages and other Web content to a user based on the geographic locations of the user, the origin of the webpage and a content delivery server. - Webopedia*

So not only does a CDN save bandwidth costs from your hosting provider, but by serving information it helps to increase the performance of your website, which has the potential to increase your search results as well as improving the user experience.

NOTES

_____

_____

_____

_____

_____

# Why should you use a CDN?

There are three basic reasons for using a CDN:

1. **Speed**
   this isn't always easy to see when you have one image on a blog post, but if you have a portfolio page on a CDN vs. your WordPress media file, you will be amazed at how much faster the page loads.

2. **User Experience**
   this has a lot to do with speed. The less time a visitor has to wait for a page to load, the more likely they are to stay on your site, so your bounce rate will go down.

3. **Improved SERP**
   again this has to do with the speed of your site. Google has stated that faster sites tend to rank higher in search results.

# What can you put on a CDN?

You can use a CDN for static content which includes images, pdf files, stylesheets (css files), javascripts, etc. A static file is a file that doesn't change. They are uploaded to your server and remain in the same state unless you download them to modify and replace them.

Other ways to minimize impact of images:
- ☐ Image Cleanup
- ☐ Resize Image After Upload
- ☐ WP Smush.it

NOTES

_____

_____

_____

_____

_____

_____

**TinyPng**
https://tinypng.com/
How to use with Photoshop images:
Use Save for Web to export your images as 24-bit transparent PNG files and upload them to TinyPNG.

# Optimization Tools
Performance Optimization: Order Styles and Javascript

You may have plugins that are contributing to a slowdown on your website. See which plugins might be slowing down your website.
- ☐ Debug Objects
- ☐ P3 (Plugin Performance Profiler)

### MONITORING TOOLS
- ☐ Google Pagespeed Insights for WordPress

### MOVE SCRIPTS
- ☐ Scripts to Footer

---

*The following pages offer code for your .htaccess file as well as some plugins you can use to optimize your site. It is highly recommended that you consult a professional before you modify your .htaccess file, as you could "break" your site. In addition, while these are suggestions, every site is unique and you may not need some, or any of these items. Also, this code and/or these plugins may cause conflicts with your other plugins so proceed with caution.*

---

NOTES

_____

_____

_____

_____

_____

_____

*Always make a copy of ANY file you are modifying before you make any changes and save it in its original state. You can copy and paste this back or upload it to override any changes you made. Take it from someone who has learned this lesson - from making this mistake.*

## Expires Headers

Expires headers are a way of setting long cache expirations on your static files (CSS, images, JavaScript, etc.) which can help improve performance. This is the code to add to your .htaccess file.

```
# Create expires headers
<FilesMatch "\.(ico|jpg|jpeg|png|gif|js|css|swf)$">
ExpiresActive on
ExpiresDefault "access plus 30 days"
Header unset ETag
FileETag None
</FilesMatch>
# End Create expires headers
```

NOTES

_____

_____

_____

_____

_____

_____

# Google Page Optimization

Before you update your .htaccess with the Google Page Speed code below, check with your host on what PHP version you are using and if you can use the Google Page Speed.

```
<IfModule pagespeed_module>
    ModPagespeed on
    ModPagespeedEnableFilters remove_comments,rewrite_javascript,rewrite_css,rewrite_images
    ModPagespeedEnableFilters elide_attributes,defer_javascript,move_css_to_head
    ModPagespeedJpegRecompressionQuality -1
</IfModule>
```

# Browser Caching

Place this code at the top of your htacess file.

```
## EXPIRES CACHING ##
<IfModule mod_expires.c>
ExpiresActive On
ExpiresByType image/jpg "access 1 year"
ExpiresByType image/jpeg "access 1 year"
ExpiresByType image/gif "access 1 year"
ExpiresByType image/png "access 1 year"
ExpiresByType text/css "access 1 month"
ExpiresByType application/pdf "access 1 month"
ExpiresByType text/x-javascript "access 1 month"
ExpiresByType application/x-shockwave-flash "access 1 month"
ExpiresByType image/x-icon "access 1 year"
ExpiresDefault "access 2 days"
</IfModule>
## EXPIRES CACHING ##
```

NOTES

_____

_____

_____

_____

_____

_____

# GZip Compression

This needs to be placed at the **bottom** of your .htaccess file.  Once you update your file, you can test your compression results at:

☐ http://99webtools.com/http_compression.php
☐ http://checkgzipcompression.com/.
☐ http://www.gziptest.com/

```
# compress text, html, javascript, css, xml:
AddOutputFilterByType DEFLATE text/plain
AddOutputFilterByType DEFLATE text/html
AddOutputFilterByType DEFLATE text/xml
AddOutputFilterByType DEFLATE text/css
AddOutputFilterByType DEFLATE application/xml
AddOutputFilterByType DEFLATE application/xhtml+xml
AddOutputFilterByType DEFLATE application/rss+xml
AddOutputFilterByType DEFLATE application/javascript
AddOutputFilterByType DEFLATE application/x-javascript
AddType x-font/otf .otf
AddType x-font/ttf .ttf
AddType x-font/eot .eot
AddType x-font/woff .woff
AddType image/x-icon .ico
AddType image/png .png
```

# Remove query strings

NOTES

_____

_____

_____

_____

_____

_____

This needs to be placed in functions.php file before the closing PHP tag <?

```
function _remove_script_version( $src ){
        $parts = explode( '?', $src );
        return $parts[0];
}
add_filter( 'script_loader_src', '_remove_script_version', 15, 1 );
add_filter( 'style_loader_src', '_remove_script_version', 15, 1 );
```

## Plugins
- [ ] Query Monitor
- [ ] Remove Query Strings From Static Resources

# Minify Javascript and CSS

## Plugins
- [ ] Better WordPress Minify
- [ ] Simple Minify
- [ ] WP Minify

That's it for modifying the .htaccess file. If you are going to use one of the cache plugins, you may not need some or any of these modifications, because the plugin will include them.

NOTES

_____

_____

_____

_____

_____

# Using W3 Total Cache

While W3 Total Cache is a popular cache plugin for WordPress websites, there are certain themes and other plugins that it may conflict with. Depending on the conflict, you may not be able to use this plugin, and may need to find another.

NOTE: *unlocking a technology tidbit for you*

*You should have access to your .htaccess file, FTP and host provider in case you need to deactivate and delete the plugin.*

Once you have W3 Total Cache installed, you will see a new Performance section on your WordPress dashboard. This plugin is very robust, and I will only be reviewing certain settings that I have used. As I have stated above, tread lightly and document what you are doing so you know what to undo if something goes wrong. I also test my site at each section to ensure all is well. You can always go back and modify others later if you like.

Select Performance on your dashboard to see the different options within the plugin.

- Appearance
- Plugins
- Users
- Tools
- Settings
- Documentation
- SEO
- Testimonials
- SREP Settings
- Performance

NOTES

_____
_____
_____
_____
_____

The following items should be checked, and any others should be left at their default settings.

## GENERAL SETTINGS
Click on General Settings and we will begin to toggle some of the options.

**Page Cache:**
Enable Page Cache
Page Cache method - Disk enhanced.

**Minify:**
Enable minify
Minify mode set to auto
Minify cache method - Disk
HTLM minifier - Default
JS Minifier - JSMin (default)
CSS minifier - Default

**Database Cache:**
Enable database cache
Database Cache Method - Disk

**Object Cache:**
Enable object cache
Object cache method - Disk

NOTES
_____
_____
_____
_____
_____
_____

**Page Cache**

Enable page caching to decrease the response time of the site.

Page cache:   ☑ Enable
Caching pages will reduce the response time of your site and increase the scale of your web

Page cache method:   Disk: Enhanced ▼

[Save all settings] [Empty cache]

**Minify**

Reduce load time by decreasing the size and number of CSS and JS files. Automatically remove unncessary data from CSS.

Minify:   ☑ Enable
Minification can decrease file size of HTML, CSS, JS and feeds respectively by ~10% on avera

Minify mode:   ⦿ Auto ○ Manual
Select manual mode to use fields on the minify settings tab to specify files to be minified, ot

Minify cache method:   Disk ▼

HTML minifier:   Default ▼

JS minifier:   JSMin (default) ▼

CSS minifier:   Default ▼

[Save all settings] [Empty cache]

## Browser Cache:
Enable browser cache

NOTES

### Database Cache

Enable database caching to reduce post, page and feed creation time.

Database Cache:  ☑ Enable
*Caching database objects decreases the response time of your site. Best used if object caching is not p*

Database Cache Method:  Disk ▼

[ Save all settings ]  [ Empty cache ]

### Object Cache

Enable object caching to further reduce execution time for common operations.

Object Cache:  ☑ Enable
*Object caching greatly increases performance for highly dynamic sites that use the* <u>Object Cache API</u>

Object Cache Method:  Disk ▼

[ Save all settings ]  [ Empty cache ]

### Browser Cache

Reduce server load and decrease response time by using the cache available in site visitor's web browser.

Browser Cache:  ☑ Enable
*Enable HTTP compression and add headers to reduce server load and decrease file load time.*

# NOTES

_____

_____

_____

_____

_____

_____

# PAGE CACHE

## General:
Enable Cache front page
Enable Cache feeds: site, categories, tags, comments
Enable Cache requests only for [mydomain.com] site address
*(where [mydomain.com] is your domain)*
Enable Don't cache pages for logged in users

---

General

☑ Cache front page
*For many blogs this is your most visited page, it is recommended that you cache it.*

☑ Cache feeds: site, categories, tags, comments
*Even if using a feed proxy service (like FeedBurner), enabling this option is still recommended.*

☐ Cache SSL (https) requests
*Cache SSL requests (uniquely) for improved performance.*

☐ Cache URIs with query string variables
*Search result (and similar) pages will be cached if enabled.*

☐ Cache 404 (not found) pages
*Reduce server load by caching 404 pages. If the disk enhanced method of disk caching is used, 404 page*

☑ Cache requests only for bonniedupree.com site address
*Cache only requests with the same URL as the site's site address.*

☑ Don't cache pages for logged in users
*Unauthenticated users may view a cached version of the last authenticated user's view of a given page.*

☐ Don't cache pages for following user roles
*Select user roles that should not receive cached pages:*

Administrator    Editor    Author    Contributor    Subscriber

[Save all settings]

---

NOTES

# MINIFY

## General:
Enable Rewrite URL structure
Enable Minify error notification to Admin Notification

## HTML & XML:
Enable HTML minify settings
Check Inline CSS minification
Check line break removal.

General

☑ Rewrite URL structure
*If disabled, CSS and JS embeddings will use GET variables instead of "fancy" links.*

☐ Disable minify for logged in users
*Authenticated users will not receive minified pages if this option is enabled.*

Minify error notification:    Admin Notification    ▾
*Notify when minify cache creation errors occur.*

[ Save all settings ]

HTML & XML

HTML minify settings:    ☑ Enable
                         ☑ Inline CSS minification
                         ☑ Inline JS minification
                         ☐ Don't minify feeds
                         ☑ Line break removal

Ignored comment stems:

*Do not remove comments that contain these terms.*

[ Save all settings ]

## NOTES

_____

_____

_____

_____

_____

_____

## BROWSER

### General:
Enable Set Last-Modified header
Enable Set expires header
Enable Set cache control header
Enable Set Entity tag (eTag)
Enable Set W3 Total Cache header
Enable HTTP (gzip) compression

General

Specify global browser cache policy.

☑ Set Last-Modified header
Set the Last-Modified header to enable 304 Not Modified response.

☑ Set expires header
Set the expires header to encourage browser caching of files.

☑ Set cache control header
Set pragma and cache-control headers to encourage browser caching of files.

☑ Set entity tag (eTag)
Set the Etag header to encourage browser caching of files.

☑ Set W3 Total Cache header
Set this header to assist in identifying optimized files.

☑ Enable HTTP (gzip) compression
Reduce the download time for text-based files.

☐ Prevent caching of objects after settings change
Whenever settings are changed, a new query string will be generated and appended to objects allowing the new policy t

Prevent caching exception
list:

Do not add the prevent caching query string to the specified files. Supports regular expr

## NOTES

_____

_____

_____

_____

_____

_____

Once you have completed all of these settings, go back to the Performance menu.
- Click on the button at the top to EMPTY ALL CACHES.
- That's it.

# Using WP Super Cache

Once you install WP Super Cache, you will need to activate the plugin in order to set it up.

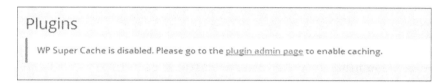

Turn Caching on and update.

Now you can move on to the Advanced tab.

NOTES

_____

_____

_____

_____

_____

_____

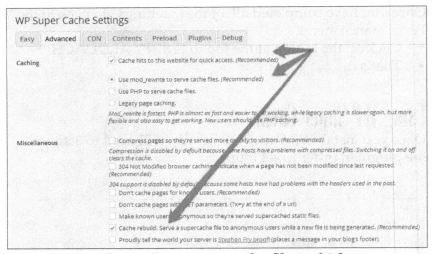

Select **Use mod_rewrite to serve cache files,** which as you can see is recommended.

Under Miscellaneous, select Cache rebuild (also recommended).

**Update status.** Scroll further and update mod_rewrite which will turn the code green, signifying you successfully updated the code.

*Make note of the uninstall directions on this page if you decide to use a different cache plugin or not use this one.*

Next you should set up a time for expiry and garbage collection. Read the content at the bottom of the page for an explanation and set a time for expiry.

NOTES

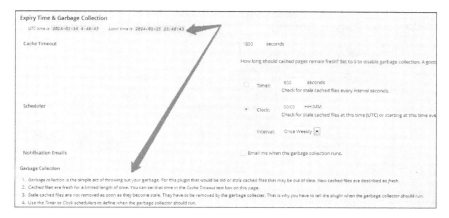

If you are using a CDN, follow the directions to enable WP Super Cache with a CDN.

Preload would also help your website optimization, but your settings would depend on your particular website. Follow the directions on the page.

NOTES

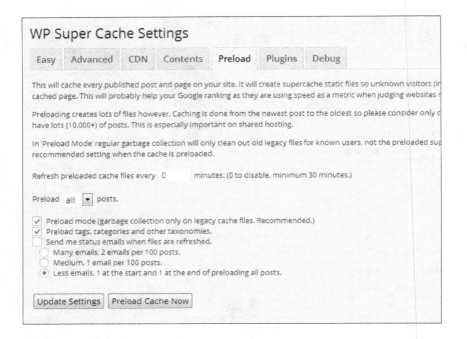

*Some host companies offer installation directions for these plugins
using their host. Check if yours does - for example:*

- *https://support.hostgator.com/articles/specialized-help/technical/
  wordpress/wp-super-cache-plugin*
- *http://www.bluehost.com/blog/educational/websites/the-ultimate-
  guide-to-supercharging-your-wordpress-blog-2097/*

NOTES

_____

_____

_____

_____

_____

_____

# Backup & Security

Learn the importance of backing up your website and integrating some security measures.

- 📖 Backing up your website
- 📖 Integrating security measures

## Backing up your website

Many businesses are so involved in setting up their websites, analytics and SEO, that they forget about backing up their site. Using a backup plugin or service is an important step in your website maintenance plan. There are usually backup plans available through your hosting provider, as well as plugins on the repository and premium plugins. Understanding your particular situation will help you determine the best solution. I cannot impress on you enough how important a backup strategy is. I have heard all too often about a website crashing, and the owner in a panic because they never backed it up.

As you develop your backup plan, you need to understand the difference between a site backup and a database backup. Your

NOTES

---

---

---

---

---

---

WordPress installation consists of at least:
- WordPress Core Installation
- WordPress Plugins
- WordPress Themes
- Images and Files
- JavaScript and PHP scripts, and other code files
- Additional Files and Static Web Pages

Your WordPress database includes your posts and data, but not all of the above. Most of these items are what creates your website from its design to the information on your site. So you need to backup both.

**How often should you backup?**
If you do not update your website on a regular basis, then creating daily, or even weekly backups is probably not necessary. On the other hand, if you blog frequently and add new content on a regular basis, then you should consider a daily backup.

**How many backups should you keep?**
WordPress recommends three backups in case a file becomes corrupt or damaged. Depending on your backup storage system, and the size of your backup, this can become costly, but should be thought of in the same way as an insurance policy.

**Can I automate my backup?**
This will depend on the plugin you select, but it is recommended that you automate your backups. One day of forgetting can be the day your site crashes after you spent hours updating and adding content. Your backup plan should include steps for

NOTES

_____

_____

_____

_____

_____

_____

automating and restoring your backup, as well as where your backups will be stored.

### Backup Plugins
- ☐ DBC Backup 2
- ☐ UpdraftPlus - WordPress Backup and Restoration
- ☐ XCloner - Backup and Restore
- ☐ WP-DBManager

### Premium Backup Plugins
- ☐ **Backupbuddy**
  http://ithemes.com/purchase/backupbuddy/

- ☐ **CodeGuard**
  http://blog.codeguard.com/codeguard-wordpress-backup-walkthrough/

- ☐ **ManageWP**
  https://managewp.com/

- ☐ **VaultPress**
  http://vaultpress.com/

- ☐ **WPMU Dev Snapshot**
  https://premium.wpmudev.org/project/snapshot/

# Integrating Security Measures

Even though you are taking steps to backup your website, you also need to implement security measures to protect against

NOTES

_____

_____

_____

_____

_____

_____

hackers. These measures range from spam protection to login restriction to strong security programs.

## Security Measures

1. Make sure that your host is backing up your website. Even though you will be backing up your website, having a website backup on your host server can expedite a restore.

2. Ensure that you have regularly scheduled backups of your website and test a restore periodically to make sure that these are working. You should also be sure that you have the ability to manually backup your site before you update any plugins or themes, and of course a WordPress update.

3. Use strong passwords. We have already discussed this earlier in this book, but it bears repeating. Do not repeat your passwords. Create unique, strong passwords for your website.

4. Do not use Admin. Again, we discussed this earlier in this book. If you use Admin for your user name, you have given the hackers 50% of the information they need to access your site.

5. Delete or rename install.php, upgrade.php and readme. html files from your WordPress installation. Once you have installed WordPress, these files are no longer needed.

6. Always deactivate unused plugins and then delete them.

NOTES

_____

_____

_____

_____

_____

Don't keep unused plugins on your site. If you don't need them, delete them.

7.  Activate Akismet or use another SPAM protection plugin. Delete SPAM regularly.

8.  Clean up your site of unused plugins, post revisions, spam and database overhead. This can also help to increase your website performance.

## Your .htaccess File

You can place several security measures in your .htaccess file. Be aware that modifying your .htaccess file can also "break" your site if not done properly. So be sure to backup your site before you modify the file and have access to your FTP or host provider to restore the file in case something goes wrong. There are also plugins on page 218 that offer .htacess protection.

**DISABLE HOTLINKING**
*Hotlinking is when other website owners use your images and/or videos by hotlinking to them via the link on your website which can put a strain on your website server and bandwidth.*

```
RewriteEngine on
RewriteCond %{HTTP_REFERER} !^$
RewriteCond %{HTTP_REFERER} !^http(s)?://(www\.)?YourDomain [NC]
RewriteRule \.(jpg|jpeg|png|gif)$ - [NC,F,L]
```

NOTES

_____

_____

_____

_____

_____

_____

**DISABLE BROWSING**
*Similar to hotlinking, eliminate the ability for visitors to browse your WordPress directories.*

```
# disable directory browsing
Options All -Indexes
```

# Site Malware

Malware can not only wreak havoc on your website, but can also cause it to be shutdown (*at least temporarily*) by Google and others, as well as your host provider. While you may be using malware protection on your computer, this usually only protects your computer and not your WordPress website installation.

## Malware Plugins

☐ Anti-Malware (Get Off Malicious Scripts)
☐ Sucuri Security - SiteCheck Malware Scanner
☐ Wordfence Security

# Website Security

There are several plugins that offer a variety of security protections from SQL Injection hackings, account login and .htaccess protection so that you do not have to modify your .htaccess file.

☐ All In One WP Security & Firewall
☐ Better WP Security
☐ BulletProof Security
☐ Wordfence Security

NOTES

_____

_____

_____

_____

_____

_____

# Premium Security Monitoring

There are several premium security plugins for WordPress that offer additional features. Here are some of them:

☐ **Acunetix WordPress Security Plugin**
https://www.acunetix.com/websitesecurity/wordpress-security-plugin/

☐ **Sucuri**
http://sucuri.net/wordpress-security-monitoring

☐ **Wordfence**
http://www.wordfence.com/

# Personal Security Monitoring

As a real estate professional, you probably don't spend much time in the office. Instead, you are probably at a coffeehouse, sandwich shop, or some other public space. You should never access a WiFi hotspot through anything other than a secure VPN.

**PERSONAL VPN - APPLE**
☐ **Cloak**
https://www.getcloak.com

**PERSONAL VPN - ANDROID**
☐ **StrongVPN**
http://strongvpn.com/

NOTES

_____

_____

_____

_____

_____

_____

☐ **Private Internet Access**
https://www.privateinternetaccess.com/

☐ **VyprVPN**
http://www.goldenfrog.com/vyprvpn

### FIREWALL PROTECTION
*Since you are probably using your phone over wifi networks, you should have firewall protection on your smartphone and tablets similar to the one you use on your laptop and other computers.*

☐ **Comodo**
http://personalfirewall.comodo.com/

☐ **Eset Smart Security**
http://www.eset.com/us/

NOTES

_____

_____

_____

_____

_____

_____

# Index

## A

## B

## C

# G

# H

# I

# J

# K

# P

# Q

# R

# S

# Website References

# P

# Q

# R

# S

# Z

Zoho

# WordPress Plugins

## Flickr Plugins

## Form Plugins

# G

## Google Plus

# I

## Image Reduction

# L

## Landing Pages

## Lead Capture

## LinkedIn

# M

## Malware Plugins

## Maps

## Marketing Automation Plugins

## Minify

## Monitoring Tools

## Move scripts

# P

## Page Builder

## Pinterest

# About The Author

Marna Friedman is a Business Strategist helping individuals and businesses improve their productivity by leveraging technology. She has implemented numerous systems throughout several industries that have resulted in better efficiency and profitability. Her experiences led to the development of a series of books offering tips, resources and more to help businesses. She can often be found speaking at events or teaching workshops, especially on WordPress.

Most recently, Marna took the real estate prelicense class to educate herself on what "right looks like" to complete her books for real estate professionals on how to leverage technology and use WordPress. As a result of that class, and some prompting from others, she decided to take the state licensing exam and is now a licensed real estate agent in Georgia.

She holds a BS in Business Administration and an MA in Media Studies/Communications, as well as certificates in Web Design, Web Development and Graphics/Animation. She is listed in Who's Who of American Women and Who's Who Among Emerging Leaders in America.


Marna Friedman, MA
http://mwfmedia.com
marna@mwfmedia.com